Agustín Gutiérrez and Santiago Marco (Eds.)

Biologically Inspired Signal Processing for Chemical Sensing

T0180994

Studies in Computational Intelligence, Volume 188

Editor-in-Chief

Prof. Janusz Kacprzyk
Systems Research Institute
Polish Academy of Sciences
ul. Newelska 6
01-447 Warsaw
Poland
E-mail: kacprzyk@ibspan.waw.pl

Agustín Gutiérrez
Santiago Marco
(Eds.)

Biologically Inspired Signal Processing for Chemical Sensing

 Springer

Dr. Agustín Gutiérrez
Departament d'Electronica
Universitat de Barcelona
Martí i Franqués 1
08028-Barcelona
Spain
Email: agutierrez@el.ub.es

Dr. Santiago Marco
Departament d'Electronica
Universitat de Barcelona
Martí i Franqués 1
08028-Barcelona
Spain
Email: santi@el.ub.es

ISBN 978-3-642-10121-2   e-ISBN 978-3-642-00176-5

DOI 10.1007/978-3-642-00176-5

Studies in Computational Intelligence   ISSN 1860949X

Typeset & Cover Design: Scientific Publishing Services Pvt. Ltd., Chennai, India.

Printed in acid-free paper

9 8 7 6 5 4 3 2 1

springer.com

Preface

Biologically inspired approaches for artificial sensing have been extensively applied to different sensory modalities over the last decades and chemical senses have been no exception. The olfactory system, and the gustatory system to a minor extent, has been regarded as a model for the development of new artificial chemical sensing systems. One of the main contributions to this field was done by Persaud and Dodd in 1982 when they proposed a system based on an array of broad-selective chemical sensors coupled with a pattern recognition engine. The array aimed at mimicking the sensing strategy followed by the olfactory system where a population of broad-selective olfactory receptor neurons encodes for chemical information as patterns of activity across the neuron population. The pattern recognition engine proposed was not based on bio-inspired but on statistical methods. This influential work gave rise to a new line of research where this paradigm has been used to build chemical sensing instruments applied to a wide range of odor detection problems.

More recently, some researchers have proposed to extend the biological inspiration of this system also to the processing of the sensor array signals. This has been motivated in part by the increasing body of knowledge available on biological olfaction, which has become in the last decade a focus of attention of the experimental neuroscience community. The olfactory system performs a number of signal processing functions such as preprocessing, dimensionality reduction, contrast enhancement, and classification. By mimicking the olfactory system architecture using mathematical models, some of these processing functions have been applied to arrays of broad-selective chemical sensors.

The latests advances in this area where presented in the GOSPEL Workshop on Bio-inspired Signal Processing held in Barcelona 2007. This workshop gathered for the first time researchers working on bio-inspired processing for chemical sensing from around the world. One of the outcomes of this workshop was the project of bringing together research contributions of this field in a book. This volume is composed of extended versions of some contributions to the workshop plus some additional contributions from other experts in the field.

The book is organized in two sections: biological olfaction; and artificial olfaction and gustation. The first section focuses on the study and modeling of the processing functions of the olfactory system. In Chapter 1, the author revises the insect olfactory system from an information processing point of view. In Chapter 2, a signal processing architecture based on the mammalian cortex is proposed. In Chapter 3, the author presents an experimental work to understand the high sensitivity of insects. In Chapter 4, the authors have performed non-invasive recordings of the olfactory bulb activity and present a technique to analyze the chemical information on these recordings. The second section is devoted to bio-inspired approaches to process chemical sensor signals. In Chapter 5, the authors propose a sensor chamber based on the olfactory mucosa that improves odor separation through temporal dynamics. In Chapter 6, the authors use a model of olfactory receptor neuron convergence to improve the correlation between sensor responses to an odor and his organoleptic properties. Chapter 7, the authors propose a method to convert chemical sensor signals to spike trains along with the processing of the signals based on the receptor neurons convergence. Chapter 8, the authors analyze the signal processing needs of an artificial chemical sensing system to detect malodors in open environments. In Chapter 9, the authors propose a chemical detection system for chemicals in liquid solution based on voltametric sensors.

We would like to thank the authors of this volume and the reviewers that helped to improve the quality of the chapters. We are also grateful to Springer's editorial staff, in particular to Professor Janus Kacprzyk that encouraged us to produce this scientific work. We also like to thank the network of excellence GOSPEL FP6-IST 507610 for its support in organizing the Workshop on Bioinspired Signal Processing. We hope that the reader will share our excitement on this volume and will find it useful.

Barcelona,   Agustín Gutiérrez-Gálvez
December 2008   Santiago Marco

Contents

List of Contributors

Alegret S.
Sensors & Biosensors Group,
Department of Chemistry, Autonomous
University of Barcelona,
Bellaterra, Catalonia, Spain

Simon Benjaminsson
Royal Institute of Technology (KTH),
Dept Computational
Biology
Stockholm, Sweden
simonbe@csc.kth.se

Maurizio Canosa
Department of Electronic Engineering,
University of Rome
"Tor Vergata"via del Politecnico
1; 00133 Roma, Italy

Raul Cartas
Sensors & Biosensors Group,
Department of Chemistry, Autonomous
University of Barcelona,
Bellaterra, Catalonia, Spain
Raul.Cartas@campus.uab.es

Arnaldo D'Amico
Department of Electronic Engineering,
University of Rome "Tor Vergata"via del
Politecnico 1; 00133 Roma, Italy

Corrado Di Natale
Department of Electronic Engineering,
University of Rome "Tor Vergata"via del
Politecnico 1; 00133 Roma, Italy

Francesca Dini
Department of Electronic Engineering,
University of Rome
"Tor Vergata"via del Politecnico 1;
00133 Roma, Italy

Julian W. Gardner
School of Engineering,
University of Warwick,
Coventry CV4 7AL,
United Kingdom

Gutiérrez J.M.
Bioelectronics Section,
Department of Electrical Engineering,
CINVESTAV, Mexico City, Mexico

Ricardo Gutierrez-Osuna
Department of Computer Science,
Texas A&M University,
520 Harvey R. Bright Bldg,
College Station, TX 77843-3112, USA
rgutier@cs.tamu.edu

Hernandez P.R.
Bioelectronics Section,
Department of Electrical Engineering,
CINVESTAV, Mexico City, Mexico

Christopher Johansson
Royal Institute of Technology (KTH),
Dept Computational Biology
Stockholm, Sweden
cjo@csc.kth.se

Karl-Ernst Kaissling
Max-Planck-Institut fuer
Verhaltensphysiologie.
Seewiesen, 82319 Starnberg,
Germany
Kaissling@orn.mpg.de

Anders Lansner
Royal Institute of Technology (KTH),
Dept Computational Biology
Stockholm University, Dept Numerical
Analysis and Computer Science,
AlbaNova University Center SE - 106 91
Stockholm, Sweden
ala@csc.kth.se

Leija L.
Bioelectronics Section,
Department of Electrical Engineering,
CINVESTAV, Mexico City, Mexico

Manel del Valle
Sensors & Biosensors Group,
Department of Chemistry,
Autonomous University of Barcelona,
Bellaterra, Catalonia, Spain

Santiago Marco
Departament d'Electrònica.
Universitat de Barcelona.
Martí i Franquès, 1.
E 08028 Barcelona
Spain

Eugenio Martinelli
Department of Electronic Engineering,
University of Rome
"Tor Vergata"via del Politecnico 1;
00133 Roma, Italy

Merkoçi A.
Sensors & Biosensors Group,
Department of Chemistry,
Autonomous University of Barcelona,
Bellaterra, Catalonia, Spain

Ivan Montoliu
Departament d'Electrònica.
Universitat de Barcelona.
Martí i Franquès, 1.
E 08028 Barcelona, Spain

Moreno-Barón L.
Sensors & Biosensors Group,
Department of Chemistry,
Autonomous University of Barcelona,
Bellaterra, Catalonia, Spain

Muñoz R.
Bioelectronics Section,
Department of Electrical Engineering,
CINVESTAV, Mexico City, Mexico

Jacques Nicolas
Department of Environmental
Sciences and management,
Monitoring unit,
University of Liège,
Avenue de Longwy 185, 6700
Arlon, Belgium

Thomas Nowotny
Centre for Computationa
Neuroscience and Robotics,
University of Sussex,
Falmer, Brighton BN1 9QJ, UK,
United Kingdom
T.Nowotny@sussex.ac.uk

Tim C. Pearce
Department of Engineering,
University of Leicester,
Leicester LE1 7RH,
United Kingdom

Giorgio Pennazza
Department of Electronic Engineering,
University of Rome
"Tor Vergata"via del Politecnico 1;
00133 Roma, Italy

Krishna C. Persaud
SCEAS. The University of Manchester.
P.O. Box 88, Sackville St. Manchester
M60 1QD
United Kingdom

Baranidharan Raman
Department of Computer Science,
Texas A&M University,
520 Harvey R. Bright Bldg,
College Station, TX 77843-3112, USA
Process Sensing Group,
Chemical Science and
Technology Laboratory,
National Institute of Standards and
Technology (NIST),
100 Bureau Drive
MS8362, Gaithersburg, MD
20899-8362, USA
baranidharan.raman@nist.gov

Anne-Claude Romain
Department of Environmental Sciences
and management,
Monitoring unit,
University of Liège,
Avenue de Longwy 185,
6700 Arlon, Belgium,
acromain@ulg.ac.be

Manuel A. Sánchez-Montañés
Escuela Politécnica Superior,
Universidad Autónoma
de Madrid,
28049, Madrid, Spain

M. Shah
SCEAS. The University of Manchester.
P.O. Box 88, Sackville St. Manchester
M60 1QD
United Kingdom

Biological Olfaction

Part I

Biological Olfaction

1

"Sloppy Engineering" and the Olfactory System of Insects

Thomas Nowotny

Centre for Computational Neuroscience and Robotics, University of Sussex,
Falmer, Brighton BN1 9QJ, UK
T.Nowotny@sussex.ac.uk

Abstract. Research on nervous systems has had an important influence on new information processing paradigms and has led to the invention of artificial neural networks in the past. In recent work we have analyzed the olfactory pathway of insects as a pattern classification device with unstructured (random) connectivity. In this chapter I will review these and related results and discuss the implications for applications in artificial olfaction. As we will see, successful classification depends on appropriate connectivity degrees and activation thresholds, as well as large enough numbers of neurons, because the strategy essentially rests on the law of large numbers. Taken as an inspiration for artificial olfaction, the analysis suggests a new paradigm of random kernel methods for odour and general pattern classification.

1.1 Introduction

Olfactory space is structurally very different to other sensory spaces. Unlike vision, which has a clear two-dimensional spatial structure of sensory input on the retina or audition which has a one-dimensional mapping of frequency space in the cochlea, odor space has no obvious neighbourhood structure. Certainly, odorants, that is mixtures of volatile chemicals, can be organized by various similarity measures. For organic compounds, by far the most frequently encountered odorants, one can compare by functional groups, carbon chain length or general molecular size. It is, however, not clear how to organize these different properties in an overall odor space. The fact that many odors humans perceive as elemental are actually mixtures of many different volatiles complicates the picture even more. The smell of a rose, for example, has dozens of different components.

Due to the complex structure of odor space, it is an extremely interesting and challenging question whether the olfactory system has a correlate of retinotopic maps in vision or frequency maps in audition. And if it does, what would be the organizing principle of such an odor "map"? Furthermore, can we learn from this organization of biological olfactory systems to build artificial chemosensor systems that perform at levels comparable to the performance of the former in general olfactory sensing tasks?

A word of caution might be appropriate at this point. For many applications, the performance of single chemical sensors, i.e., sensors that specifically respond to certain chemical substances or general analysis techniques like mass spectrometry and

A. Gutiérrez and S. Marco (Eds.): Biologically Inspired Signal Processing, SCI 188, pp. 3–32.
springerlink.com   © Springer-Verlag Berlin Heidelberg 2009

gas chromatography are vastly superior to biological systems in particular in analyzing the components of a chemical mixture or finding small traces of a specific chemical. However, whenever human sensory impressions are the target, e.g., in assessing whether the odor in a print shop or close to a landfill are acceptable to humans or not (see xxx in this volume), or, whether a product smells or tastes right, one clearly will need a more bio-mimetic approach. Another aspect is the "online" use in a complex environment. Examples of applications are searching for drugs or explosives on a busy airport or for truffles in the forest. For the former it might appear easy to look for suspicious Nitrogen peaks in a mass-spectrograph. The number of possible false positives is, however, very large.

In this chapter I will review our recent work on the olfactory system of insects, in particular the locust. The main hypothesis underlying this work is that a generally unstructured, random connectivity may be sufficient, if not advantageous, for classifying odors. I will start by reviewing the known anatomical, bio-chemical and electro-physiological properties of the olfactory system and then embark on a general analysis of a corresponding model system with unstructured connectivity. This analysis is performed on several levels of description and we will see how each level of description allows different insights. After discussing aspects of temporal coding which is believed to be of great importance in this system I will conclude with a general discussion in particular with a focus on the question of applicability of this theoretical work for building artificial devices.

1.1.1 The Olfactory Pathway of Insects

Insects smell with their antennae. While the olfactory systems of different insects bear many similarities, they are by no means identical. For sake of concreteness I will here mainly focus on the anatomy and electro-physiology of locust (*Schistocerca Americana*). This species has been studied for a long time, in particular in the group of Gilles Laurent at the California Institute of Technology. The locust has on the order of 50,000-100,000 olfactory receptor neurons (ORN) on either of its antennae (Ernst 1977, Hansson 1996). The ORN express olfactory receptor proteins that bind to the volatiles carried to the antennae through the air. Each receptor type typically responds to many different chemicals and each chemical typically can bind to several different receptors. Even though it has not been shown in locust, each ORN likely expresses only one type of olfactory receptor (OR) in analogy with findings in fruit fly (*Drosophila*) and mouse (Gao et al. 2000; Vosshall et al. 2000; Scott et al. 2001). By the same analogy, all ORN with receptors of the same type presumably converge to the same regions (Gao et al. 2000, Vosshall et al. 2000, Scott et al. 2001) in the antennal lobe (AL), the first brain structure of olfactory information processing. Whereas flies, bees and mammals have a few large regions of input convergence, the so-called (macro) glomeruli, locust have a micro-glomerular structure of many, small glomeruli in the antennal lobe (Ernst 1977). For the purpose of our analysis we contended ourselves with the knowledge that whichever the exact organization of the input convergence, in either case there are a few, in the case of locust about 800, principal neurons (PNs) which relay the odor information to higher brain structures, foremost the mushroom bodies and the lateral horn of the protocerebrum. It also appears sound to assume that each PN connects to only one (Wang et al. 2003), maybe a few (Bhalerao et al. 2003; Sato et al. 2007), glomeruli thus sampling the input of only one, at most a few, olfactory receptors. Due to this one OR - one ORN – one glomerulus – one PN organization the olfactory code on the level of the input side to the PN in the AL is about the same code of overlapping patterns of activity that the response profiles of the ORs define.

Within the AL about 300 non-spiking interneurons modulate the output of the PNs. The resulting activity has been experimentally characterized as patterned on two time scales. There are fast 20 Hz local field potential (LFP) oscillations to which spikes of PNs are locked (Laurent et al. 1996; Wehr and Laurent 1996). The PNs are active in synchronized groups and these groups evolve over time in a slower, odor-specific pattern. The slow switching dynamics has been hypothesized to improve odor discrimination for very similar odors (Laurent et al. 2001; Rabinovich et al. 2001) and some experimental evidence for a decorrelation, and therefore presumably disambiguation, of similar patterns in the olfactory bulb of zebra fish has been observed experimentally (Friedrich and Laurent 2001).

Table 1.1. Some known properties of the olfactory system of insects. For the work described here, the assumption of random connections and the localization of learning to the synapses between intrinsic and extrinsic Kenyon cells are particularly important.

Structure	Antenna	Antennal Lobe	KC / Mushroom Body / Lobes (eKC)		
Neurons	90000 ORN	800 glomeruli	800 PN	50000 KC	~ 100s eKC
connectivity	Genetically encoded		Unknown, random?		
		Genetically encoded		Unknown, random?	
Concentration dependence	dependent	Dependent	Unknown	Independent	Unknown
Plasticity	Adaptation	Plastic in odour conditioning	PN adaptation	Main locus of plasticity in odour conditioning	

Feedforward inhibition synchronous with the LFP in the AL is transmitted via an indirect pathway involving the lateral horn onto the KCs of the MB. This feedforward inhibition, combined with the presumed function of KCs as coincidence detectors, effectively partitions the slow temporal patterns of PN activity in the AL into snapshots of active PNs during each LFP cycle (Perez-Orive et al. 2002). Following this interpretation, the challenge of the MBs is to classify patterns of active PNs in each snapshot (and/or across several snapshots) and extract odor information from this code. The MBs of locust have about 50000 KCs which synapse onto a few, maybe hundreds, of output neurons in the MB lobes, so-called extrinsic Kenyon cells (eKCs). In summary, many ORNs project to a few PNs in the AL which fan out to a large number of KCs. These in turn converge onto a small number of eKCs that constitute the output of the system towards pre-motor areas. For a recent, more complete discussion of the locust anatomy and electrophysiology see (Farivar 2005).

The connectivity between PNs in the AL and the KCs, as well as between KCs and eKCs is largely unknown. Several attempts have been made, mainly using staining techniques in the fruit fly Drosophila, to elucidate whether the connectivity is stereotypic or animal-specific (Wong et al. 2002; Marin et al. 2002; Masuda-Nakagawa et al. 2005). To our mind, the results remain somewhat ambiguous and contradicting. It is obvious that the genome is not large enough to specify each single synaptic connection between the neurons of the system. Therefore, the connections are either formed by some smart targeting mechanism (like in the earlier stages from receptor neurons to glomeruli and subsequently projection neurons (Gill and Pearce 2003; Tozaki et al 2004; Feinstein et al. 2004; Feinstein and Mombaerts 2004) or they are formed generically by proximity of dendritic structures to axons, randomly in a sense.

In the following I will present results on whether odor recognition in this system can be understood based on the general morphology in terms of cell numbers and average connection degrees and the assumption of *completely unstructured* (random) connectivity between the AL and the MB. Apart from trying to understand the olfactory information processing in insects this also raises some interesting general questions. Can neural networks function simply relying on the law of large numbers? And if so, how well do they perform? And, of course, may there be implications for technical applications in this "sloppy engineering" as well?

1.1.2   Assumptions and Conventions

Throughout this chapter I will use the assumption that cells make synapses with each other independently from other cells and with fixed probabilities denoted by $P_{PN \to KC}$ and $P_{KC \to eKC}$. The analyses presented below comprise different levels of description.

To address general questions of connectivity and function, the system is modeled with time-discrete, binary neurons (McCulloch and Pitts 1943) which fire (have value 1) whenever the sum of active, connected neurons in the previous time step is larger than their firing threshold θ,

$$y_i(t+1) = \begin{cases} 1 & \sum_j b_{ij} x_j(t) > \theta \\ 0 & \text{otherwise} \end{cases} \tag{1.1}$$

Where b_{ij} encodes the synaptic connectivity which is also simplified to binary values, $b_{ij} = 1$ if j synapses onto KC i and $b_{ij} = 0$ otherwise. Similarly, we use c_{ij} to the connections from KCs to eKCs. The numbers of neurons in the different structures are denoted by N_{PN}, N_{KC} and N_{eKC}.

Throughout this chapter we will use a common notation of x_i for the AL neurons, y_i for the intrinsic KCs of the MB and z_i for the eKCs or output neurons in the MB lobes.

The biological relevance and feasibility of the results obtained with this connectionist approach are confirmed in a second stage of analysis with computer simulations of realistic, spiking neuron models presented in section 1.3. I will indicate the details of the models when they are introduced and will refer to previously published work where appropriate.

The remainder of this chapter is organized into 3 sections discussing the fan-out properties of PN to KC connections and the AL-MB system as a classification machine (section 1.2), more detailed models of the AL-MB classification system (section 1.3) and an analysis of lateral connections between KCs in the MB for time integration (section 1.4). I will close with a general discussion of the implications of these results for biomimetic approaches to odor classification and pattern classification in general.

1.2 The AL-MB Fan-Out Phase

The connections from the PNs of the AL to the KCs of the MB are strongly divergent from some hundreds to fifty thousand neurons. As we will discuss in more detail later (section 1.3), we would like to suggest that this divergent connectivity serves to separate the strongly overlapping PN activity patterns by projection into the higher-dimensional KC space. At the least, one will therefore expect that this fan-out connectivity should aim at loss-less information transmission (in contrast to fan-in or convergent connectivity where loss of information is inevitable). For simplicity, let us assume that the input patterns, i.e., the activity patterns of PN neurons within one LFP cycle in the AL, can be described as arising from each PN being active independently with a fixed probability $p_{PN} = 0.1$, which corresponds to the experimental observation that about 10% of PNs are active at any given time.

Similarly, we assume that connections between neurons are formed independently from each other with fixed probability $p_{PN \rightarrow KC}$. The value for $p_{PN \rightarrow KC}$ is not well known. The common consensus for this connection density has been a value on the order of $p_{PN \rightarrow KC} = 0.05$ or less, even though recent experiments suggest much larger values close to $p_{PN \rightarrow KC} = 0.5$ (Jortner et al. 2007). The requirement of lossless information transmission induces limits on this connection probability and the firing threshold of KCs, the other main unknown in the system. A further, general criterion for the nervous system is the minimization of energy cost (Laurent 1999) which may suggests further restrictions for suitable system parameters.

Before presenting our results on structural implications of the observed and predicted levels of activity in the MB and of the requirement of lossless information transmission, some explanation of our probabilistic approach seems warranted. The probability P_{PN} of activity in a given PN mainly reflects properties of the input space (odor space) and different patterns of PN activity are "diced out" for every LFP cycle. The connection probabilities $P_{PN \rightarrow KC}$ and $P_{KC \rightarrow eKC}$, on the other hand, refer to the random connectivity of each locust, i.e., the connectivity is determined only once for each animal. In building distributions (and taking averages) with respect to both probability spaces, we are making statements about the distribution of (and the typical value of) properties for all locusts in response to all possible odors, in a sense.

1.2.1 Activity Levels in the MB

Using the above assumptions of independently chosen random connections and independently and randomly active PNs we can directly calculate the probability for a given KC to be active,

$$p_{KC} = \sum_{k=\theta}^{N_{PN}} \binom{N_{PN}}{k} \left(p_{PN} p_{PN \rightarrow KC} \right)^k \left(1 - p_{PN} p_{PN \rightarrow KC} \right)^{N_{PN}-k} \quad (1.2)$$

Because all KCs "look" at the same PN activity patterns at any given time, their activity is not independent, such that the probability distribution for the number n_{KC} of active KCs is not a binomial distribution with parameters p_{KC} and N_{KC}. It is given by a more intricate expression derived in Appendix A. See also (Nowotny and Huerta 2003) for a more explicit derivation. The distribution of the number n_{KC} of active KCs in each LFP cycle is given by (A1) and shown in Figure 1.1A in comparison to the naïve guess of a binomial distribution with parameters N_{KC} and p_{KC}. Surprisingly, the expectation value for the number of active neurons, i.e., of the distribution (A1), is $N_{KC}p_{KC}$ as naïvely expected. The standard deviation, however, is much larger. The oscillations in the distribution are not a numerical error but due to the discrete (integer) firing thresholds of the McCulloch–Pitts neurons.

The expectation value for the number $N_{KC}p_{KC}$ of active KCs depends critically on the firing threshold θ and the connectivity probability $p_{PN \rightarrow KC}$ according to (1.2) and is shown in Figure 1.1B (black line). Not surprisingly, larger thresholds lead to more sparse activity and, conversely, more connections to higher activity levels. Thus, in terms of energy efficiency, very large thresholds and sparse connectivity are favorable.

In summary, the activity level in the MB depends essentially on the connectivity degree and the firing threshold of KCs as expected. More surprising is the extremely wide distribution of possible activity levels which may introduce the necessity of gain control mechanisms.

1.2.2 Lossless Information Transmission

As a measure for the faithfulness of the information transmission from the AL to the MB we use the probability of having the same KC activity pattern in the MB in response to two different input patterns in the PNs of the AL. Clearly, this probability has to be kept to a minimum as this situation corresponds to confusing two odors. Appendix A explains how to calculate this "probability of confusion". The results for $p_{PN \to KC} = 0.05$ are shown in Figure 1.1B (grey bars). The figure actually displays the approximation

$$P(\vec{y} = \vec{y}') = P(\vec{y} = \vec{y}' \mid \vec{x} \neq \vec{x}')P(\vec{x} \neq \vec{x}') + P(\vec{x} = \vec{x}')$$
$$\approx P(\vec{y} = \vec{y}' \mid \vec{x} \neq \vec{x}') = P_{confusion}. \tag{1.3}$$

This approximation is almost an equality because for $p_{PN} = 0.1$ we have

$P(\vec{x} = \vec{x}') = \left(p_{PN}^2 + (1 - p_{PN})^2 \right)^{N_{PN}} = 6.5 \cdot 10^{-87}$ and, therefore, almost 0, while $P(\vec{x} \neq \vec{x}') = 1 - P(\vec{x} = \vec{x}')$ is almost 1.

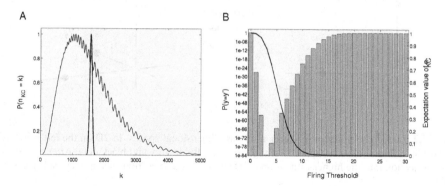

Fig. 1.1. a) Probability distribution for the number of active KCs calculated correctly (thin line, $\theta = 6$) and with the incorrect assumption of statistical independence of KCs (thick line). B) Probability of collision (bars) and expectation value for the number of active KCs (line). Note that the minimum for the collision probability is reached for fairly large numbers of active cells. On the other hand, the collision probability remains acceptably low ($< 10^{-6}$) for thresholds up to $\theta = 13$. $N_{PN} = 1000$, $N_{KC} = 50000$, $p_{PN} = 0.1$, and $p_{PN \to KC} = 0.05$ in both panels.

For very small values of $p_{PN \to KC}$ one can assume approximate independence of the inputs to the different KCs, leading to a much simplified expression (Garcia-Sanchez and Huerta 2003),

$$P(\vec{y} = \vec{y}' \mid \vec{x} \neq \vec{x}') \approx \left(p_{KC}^2 + (1 - p_{KC})^2 + 2\sigma_{KC}^2 \right)^{N_{KC}}, \tag{1.4}$$

where p_{KC} is the expected probability of a KC to be active given by (1.2) and σ_{KC} denotes the standard deviation of the distribution of single cell activity (seen as a random variable with respect to the probability space of potential connectivities).

The goal was to minimize this probability of confusion of odor input patterns. As Figure 1.1B shows, the minimum for the confusion is realized for $\theta = 3$ but at rather huge activity levels of more than 50 %. On the other hand, it may be questionable whether it is important to have a confusion probability of 10^{-84} rather than, let us say, 10^{-6}. For all practical purposes both are basically zero. We will in the following, therefore, only require that the collision probability $P(\bar{y} = \bar{y}' | \bar{x} \neq \bar{x}') \approx P(\bar{y} = \bar{y}')$ is less than a given small tolerance level. For a tolerance level of $P(\bar{y} = \bar{y}') < 10^{-6}$, the firing threshold can be $1 \leq \theta \leq 13$ (Figure 1.1B) leading to potential activity levels down to less than 1%.

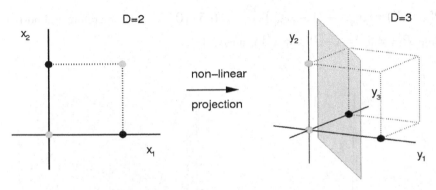

Fig. 1.2. Linear classification is easier in higher dimensional spaces. In 2D on the left it is impossible to find a linear subspace (a straight line) to divide grey and black dots. After a nonlinear projection into the higher-dimensional 3D space, it is easy to find such a linear subspace (a plane).

Another requirement for the system is that all neurons should be able to fire, i.e., all neurons should have θ or more incoming connections. The expected number of "useless neurons" with fewer connections is easily calculated to be

$$n_{\text{silent}} = N_{KC} \sum_{j=0}^{\theta-1} \binom{N_{PN}}{j} P_{PN \to KC}{}^{j} \left(1 - P_{PN \to KC}\right)^{N_{PN} - j}. \tag{1.5}$$

By requiring that no silent neurons exist, i.e., $n_{\text{silent}} < 1$, we obtain another restriction on connectivity and firing threshold. For example, if $N_{PN} = 1000$, $P_{PN \to KC} = 0.05$, and $N_{KC} = 50000$ then the threshold θ cannot be larger than 24. This condition is weaker than the previous condition derived from the "no confusion" requirement above. For other cell numbers and connectivity degrees this, however, need not be the case.

Realistically, one may also be interested in a more robust coding scheme in which the hamming distance $D_H \equiv \sum_{i=1}^{N_{KC}} |y_i - y_i'|$ is required to be greater than some minimum distance k, i.e., in which the activity patterns in the MB differ by more than k active neurons. So far, we have only considered $k = 0$. Calculating the confusion probability for the more general requirement with $k > 0$ is beyond the scope of this chapter and the interested reader is referred to (Garcia-Sanchez and Huerta 2003), at least for an approximation. It is, however, clear that $k > 0$ is a stronger condition and includes the case $k = 0$. The parameters specified with the stronger condition in place will, therefore, fall into a subset of the allowed parameter region determined by our less stringent condition with $k = 0$.

In summary, we have seen that if parameters (connectivity degree and firing threshold) are chosen wisely, fully random connections allow an almost always (in the loose sense of a very small failure probability) one-to-one projection of activity patterns from the AL to the MB, a necessary requirement for successful odor classification. At the same time, the activity level in the MB can remain reasonably low even though the absolute minimum for the confusion probability is attained at very high activity levels.

1.3 The Classification Stage

The projection from the PNs in the AL to the much larger number of KCs in the MB presumably serves the function of separating the overlapping activity patterns in the AL, that ensue from the wide response profiles of the ORNs, into non-overlapping, sparse activity patterns in the MB. These can subsequently easily be classified with linear classifiers. The idea of such a two-stage classification scheme dates back to the seminal work of Thomas Cover (Cover 1965) and has been exploited heavily in form of the so-called support vector machines (Cortes and Vapnik 1995). That principle behind the scheme is illustrated in Figure 1.2. It is important to note, that neurons are, in approximation, linear classifiers because they fire when the (almost) linear sum of their inputs exceeds a certain threshold. It is, therefore, quite natural to think of neural systems in these terms. The suggested classification scheme needs two more essential ingredients: plasticity of the synapses between intrinsic and extrinsic KCs and competition between the eKCs. The intuition behind these requirements is that the connections from KCs to eKC are initially randomly chosen, such that some eKC will receive somewhat larger input for one odor and others receive more for another odor. Having a correlation based ("Hebbian") plasticity rule for the synapses, neurons that do respond often to a certain odor will strengthen their synapses and respond even more reliably in the future. Through competition between the output neurons, the strongest responders for each given odor input prevail and others cease to respond. In this way, we have created a so-called winner-take-all situation. After some experience of odor inputs, we expect the system to respond reliably with specific neurons to specific odors.

To test this idea and the conditions on how the system needs to be organized in terms of the connectivity, initially and after learning, we introduced a "Hebbian learning rule" for the synapses from KCs to eKCs:

$$c_{ij}(t+1) = F\left(y_j, z_i, c_{ij}(t)\right) \qquad (1.6)$$

Where F is a probabilistic function,

$$F(1,1,c_{ij}) = \begin{cases} 1 & \text{with probability } p_+ \\ c_{ij} & \text{with probability } 1 - p_+ \end{cases}$$

$$F(1,0,c_{ij}) = \begin{cases} 0 & \text{with probability } p_- \\ c_{ij} & \text{with probability } 1 - p_- \end{cases} \qquad (1.7)$$

$$F(y_j, z_i, c_{ij}) = c_{ij} \text{ in all other cases}$$

If both the presynaptic and the postsynaptic neurons are active, the synapse becomes active with probability p_+ and remains unchanged with probability $1 - p_+$. This implements the Hebbian, correlation-based strengthening of synapses between neurons that are active at the same time. The probability p_+ regulates the speed with which synaptic connectivity patterns change; it is the "speed of learning", in a sense. If the presynaptic neuron is active but the postsynaptic one is not, the synapse is turned off or removed with probability p_- and remains unchanged with probability $1 - p_-$. This rule implements a part of the competition between output neurons discussed above. If a neuron fails to respond to some input it becomes even less likely to respond at a later time, conceding defeat to another neuron. The probability p_- regulates the speed of "concession of defeat". In all other cases the synaptic connections remain unaltered.

As in the treatment of the fan-out phase, we analyzed the performance of the system in two steps. First, we consider a simplified task of classifying one odor input as different from all other possible inputs. In a second step we then look at the more realistic task of separating (recognizing) different classes of odor inputs in a structured set of odor input patterns.

1.3.1 Classification of One Odor Input

In the simplified task of distinguishing one odor from all other odor input patterns only one output neuron, eKC, is required. We consider a simple learning paradigm in which the eKC is made to fire if the trained input, let us call it cherry, is present, and prevented from firing for all other randomly chosen inputs. A careful inspection of the learning equations (1.6) and (1.7) reveals that, given this protocol is repeated long enough, the connections to the eKC will converge to a connectivity pattern in which the connections to the eKC from KCs which respond to cherry are present and the connections from any other KCs are absent (unless one of the learning probabilities p_+ and p_- is zero). For a successful classification of one odor it is then sufficient that

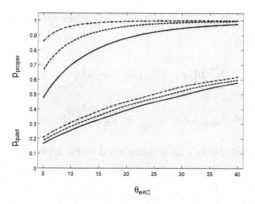

Fig. 1.3. Probability of proper classification of one odor input pattern against 10 other randomly chosen input patterns (upper traces) and probability of the failure to respond to the trained pattern (lower traces) for MBs of sizes $N_{KC} = 1000$ (solid), 2000 (dashed), and 5000 (dash-dotted) against the threshold of the extrinsic KC. The KC threshold was fixed at 5 and $p_{PN \rightarrow KC}$ adjusted such that the number of active KCs was $n_{KC} = 50$.

this connectivity allows that the eKC is exclusively active when cherry is present. In other words, we need to calculate the probability that the eKC is active in response to other inputs and control that this probability is smaller than an appropriate error tolerance.

We call the probability not to fire in response to other inputs the probability of proper classification. It depends on the number of ones in the cherry

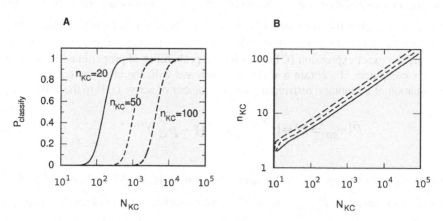

Fig. 1.4. (A) Approximated classification probability for one input against 100 other random inputs for three different numbers of active KCs and in dependence on the total number of KCs. The firing threshold of the eKC was chosen as $\theta_{eKC} = 7$. (B) Relationship of active KCs to total number of KCs defined by the crossover from failing to successful classification for 1000 (solid line), 100 (middle dashed line) and 10 (upper dashed line) competing inputs. (modified from (Huerta et al. 2004)).

KC activity pattern, l_{cherry}, and on how many other random patterns n_{pattern} we wish to classify against,

$$
P_{\text{classify}}\left(n_{\text{pattern}}\right)= \sum_{l_{\text{cherry}}=0}^{N_{\text{KC}}} P\big(z(\bar{y}_1)=0 \cap \ldots \cap z\big(\bar{y}_{n_{\text{pattern}}}\big)=0 \,|
$$

$$
l_{\text{cherry}}, \bar{y}_1 \neq \bar{y}_{\text{cherry}}, \ldots, \bar{y}_{n_{\text{pattern}}} \neq \bar{y}_{\text{cherry}} \big) P\big(n_{\text{KC}} = l_{\text{cherry}}\big).
$$

(1.8)

This conditional probability can be simplified to be approximately the probability without conditioning for $\bar{y}_i \neq \bar{y}_{\text{cherry}}$ using the same argument of almost vanishing probability $P\big(\bar{y} = \bar{y}_{\text{cherry}}\big) \approx 0$ as in equation (1.3) above.

The probability (1.8) is calculated in Appendix B. The calculation of (C1)-(C5) is computationally very expensive. We, therefore, use for the moment a smaller system of $N_{\text{PN}} = 100$ and MB sizes of $1000 \leq N_{\text{KC}} \leq 5000$. With the usual choice of activity levels in the AL, $p_{\text{PN}} = 0.1$, and connectivity $p_{\text{PN} \to \text{KC}} = 0.05$, and a moderate KC firing threshold of 5, we can adjust $p_{\text{PN} \to \text{KC}}$ such that the number of active KCs is $n_{\text{KC}} = 50$. Then we can calculate the probability of correct classification $P_{\text{classify}}\left(n_{\text{pattern}}\right)$ for $n_{\text{pattern}} = 10$ and the probability for the eKC to be quiescent (not to respond to any input, not even the trained cherry pattern). The results as a function of the eKC threshold θ_{eKC} and the MB size N_{KC} are shown in Figure 1.3. The proper classification can be achieved (P_{classify} close to one) but only with the price of non-negligible probability of quiescence. The situation clearly improves with MB size. For larger MB sizes than 5000 it is, unfortunately, almost impossible to evaluate the exact expression (C1)-(C5) for the probability of proper classification on today's computers. To obtain a wider overview we will, therefore, resort to the approximation of a binomial distribution for the number of active cells in the MB,

$$
\tilde{P}\big(n_{KC} = k\big) \equiv \binom{N_{\text{KC}}}{k} p_{\text{KC}}{}^k \big(1 - p_{\text{KC}}\big)^{N_{\text{KC}}-k},
$$

(1.9)

Where p_{KC} is given by (1.2). This approximation for (A1)-(A3) is then used in (C1) and (C3) to determine P_{classify}. With this approximation, we can obtain the dependence of the probability of proper classification on the number of KCs in the MB in a much wider region. Of course, strictly speaking, the probability of proper classification depends on several other factors as well (as we have already seen above): The size of the AL, the activation probability in the AL, the connection probability from the AL to the MB, the firing threshold of the KCs, the firing threshold of the eKC, and the number of other inputs the trained input pattern is compared against. With the simplified assumption of a binomial distribution of active KCs the dependence on the first four factors is subsumed into the expected value for active KC expressed by the KC firing probability p_{KC}. If we fix the threshold of the eKC, then, in order to have a meaningful comparison, we will adjust p_{KC} to give the same number n_{KC} of active KCs: $p_{KC}(N_{KC}) = n_{KC}/N_{KC}$. The corresponding results for $P_{classify}$ in a comparison against 100 other randomly chosen input patterns are shown in Figure 1.4. The probability of correct classification is a sigmoid function of the MB size and has a rather rapid transition from zero to one which occurs the sooner, the smaller the number of active KCs is. The transition point $P_{classify} = 0.5$ defines a relationship between the number of active KC, n_{KC}, and the total number of KC, N_{KC}, which turns out to be a power law, $n_{KC} \sim N_{KC}^{0.472}$, which origin is to date still an open question. It is worth noting that the deviation of the exponent from 0.5 is not a numerical error. The scaling, furthermore, persists for different numbers of competing inputs (Figure 1.4B).

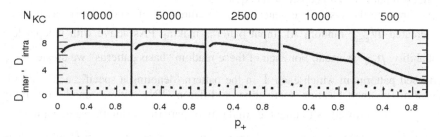

Fig. 1.5. Inter (solid line) and intra (dotted line) distances for systems of different MB size. The number of active KC was fixed to be $n_{KC} = 35$. Clearly, for MB sizes $N_{KC} > 2500$, the system is successful and then degrades for smaller MBs. The number of 2500 KCs is about the estimated size in *Drosophila*. (Modified from (Huerta at al. 2004)).

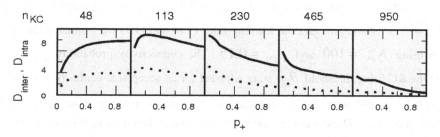

Fig. 1.6. Inter (solid line) and intra (dotted line) distances of output patterns for different levels of activity in the MB. Here, the MB size was fixed to $N_{KC} = 50000$ and n_{KC} was varied from 48 to 950 (left to right). The level of 113 active KCs seems to be optimal. For some activity levels the classification success also depends rather strongly on the learning rate p_+. (Modified from (Huerta et al. 2004)).

In summary, we have seen that adequate MB size and, most importantly, sparse activity in the KCs of the MB (small n_{KC}) are the essential ingredients for successful classification. The requirement of sparseness in the MB activity is important for the classification stage from the MB to the MB lobes and apparently supersedes the requirement of minimizing the probability of confusion in the projections from the AL to the MB. As we have already hinted above, a reasonably small probability of confusion at low activity levels seems superior to a minimized confusion at high levels of activity.

1.3.2 Classification of a Structured Set of Odor Input Patterns

The computational complexity of calculating exact, or even approximated, probability distributions for the full system with several output neurons is too daunting and we will have to resort to direct simulations from now on. To implement the competition between eKC (output) neurons, we enforce a strict winner-take-all rule in the eKCs. Instead of a fixed threshold that determines firing, we let the n_{out} neurons with the strongest input fire and keep all others quiescent.

As inputs to the system we generated a structured set of input patterns as follows. We created N_{class} random AL input patterns making PNs active with independent probability p_{PN} as usual. For each of these random "basis patterns" we create n_{class} perturbed patterns in which each 1 in the pattern (denoting a specific active PN) is moved to another location (activating a different PN) with a small probability $p_{relocate}$. This allowed us to tune the similarity of patterns within the same input class (with the same "basis pattern") by changing $p_{relocate}$. For $p_{relocate} = 0$ all the input patterns of the same class are identical and they become increasingly different from each other with increasing $p_{relocate}$.

We used a set of input patterns with $N_{class} = 40$ input pattern classes of each $n_{class} = 10$ inputs generated with $p_{relocate} = 0.1$. With the competition rule of $n_{out} = 5$ active outputs and the learning rules (1.6) and (1.7) we simulated the system using $N_{PN} = 100$ and $p_{PN} = 0.15$. The connectivity probability $p_{PN \to KC}$ and the KC firing threshold θ_{KC} were adjusted for an acceptable confusion probability ($< 10^{-5}$) and such that the expectation value for the number of active KC is a fixed value n_{KC}. These numbers reflect to some extent the characteristics of the Drosophila olfactory system but were mainly chosen due to numerical limitations that do not allow calculations at locust sizes. For the (experimentally unknown) number of eKCs we assumed $N_{eKC} = 100$ and the depression learning rate was fixed at $p_- = 0.5$ while the potentiation learning rate p_+ was used as a free parameter.

In the training phase we presented 8000 inputs from the input set in random order and then characterized the resulting output patterns in the eKCs.

As a measure for the structure of the set of observed activity patterns, be it on the input side in form of the generated set of input patterns, or on the output side in form of the set of activity patterns in response to these inputs, we used the following distance measures.

The *average intra-distance* is the average absolute distance between all patterns belonging to the same class, then averaged over all classes,

$$D_{\text{intra}}\left(\left\{\bar{x}_j^{\,i}\right\}\right) = \frac{1}{N_{\text{class}}} \sum_{i=1}^{N_{\text{class}}} \frac{1}{n_{\text{class}}} \sum_{j=1}^{n_{\text{class}}} \left\| x_j^{\,i} - \left\langle x_j^{\,i} \right\rangle_j \right\|, \tag{1.10}$$

where the upper index i labels the classes and the lower index j the specific input patterns within each class. The *inter-distance* is the average distance of the mean patterns of each class,

$$D_{\text{inter}}\left(\left\{\bar{x}_j^{\,i}\right\}\right) = \frac{1}{N_{\text{class}}(N_{\text{class}}-1)} \sum_{i=1}^{N_{\text{class}}} \sum_{k=i+1}^{N_{\text{class}}} \left\| \left\langle x_j^{\,k} \right\rangle_j - \left\langle x_j^{\,i} \right\rangle_j \right\|. \tag{1.11}$$

We performed two types of experiments. In one we kept the number of active KCs fixed at $n_{\text{KC}} = 35$ (by appropriate adjustment of $p_{\text{PN} \to \text{KC}}$ and θ_{KC}) and varied the total number N_{KC} of KCs. In the other we kept the total number $N_{\text{KC}} = 2500$ constant and varied the number n_{KC} of active KCs instead. The results are illustrated in Figures 1.5 and 1.6. Several observations deserve attention. Firstly, the same rule of thumb applies to the size of the MB as before, the bigger the better. In addition, the learning rate p_+ matters less and less the larger the number of KCs. For the number of active KCs with fixed total number of KCs (Figure 1.6) the situation is slightly different. Here, an intermediate number of active KCs, $n_{\text{KC}} = 113$ seems optimal and the learning speed always appears to matter in this situation. One tentative explanation, why the optimal level of activity in the MB is moved away from the sparsest activities here, is that in these simulations now two failure modes are possible. A neuron can respond to incorrect inputs (this is the failure mode we considered so far and that warrants extremely sparse activity levels) or it can fail to respond to an odor input it should be responding to (which warrants not too sparse activity levels).

We have seen in this subsection once again that one of the determining factors in making a system successful in the information processing framework with disordered (random) connections is the correct balance of system size, connectivity degrees and firing thresholds. Other factors like learning rates and output redundancy may play equally important roles.

1.4 More Detailed Models

In the description with McCulloch-Pitts neurons, used in the preceding sections, some
important questions had to be left unanswered due to the discrete time structure of
these models. Most importantly, the question of how the winner-take-all situation in
the MB lobes may be realized in a realistic neuronal network and how this particular
implementation determines the output code had to remain open.

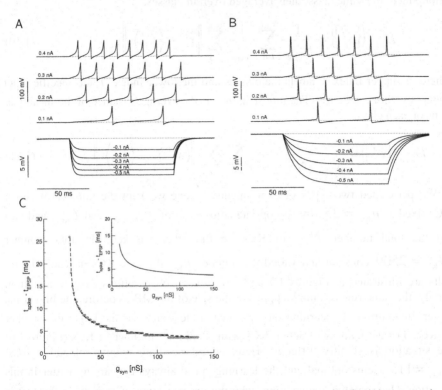

Fig. 1.7. Comparison of the map based neurons (defined by equation (1.15)) (A) to an equivalent Hodgkin-Huxley type model (details in (Nowotny et al. 2005)) (B). The response time to
an EPSP (which is the most important parameter within this context) is almost identical for
both (C). The inset shows the corresponding response curve of the honeybee KC model of
(Wüstenberg et al. 2004) which is also virtually identical. (Modified from (Nowotny et al.
2005)).

To address these questions and prove the validity of our analysis for spiking neuron models we have built a more realistic network model of an insect olfactory system
using map-based spiking neurons (Cazelles et al. 2001; Rulkov 2002). These neurons
are defined by a simple iterative map,

$$V(t + \Delta t) = \begin{cases} V_{\text{spike}}\left(\dfrac{\alpha V_{\text{spike}}}{V_{\text{spike}} - V(t) - \beta I_{\text{syn}}} + \gamma\right) & V(t) \le 0 \\ V_{\text{spike}}(\alpha + \gamma) & \begin{aligned}&V(t) \le V_{\text{spike}}(\alpha + \gamma) \\ &\& V(t - \Delta t) \le 0\end{aligned} \\ -V_{\text{spike}} & \text{otherwise} \end{cases} \quad (1.12)$$

where $V_{\text{spike}} = 60\,mV$, $\alpha = 3$, $\beta = 2.64\,M\Omega$, and $\gamma = -2.468$. The parameter β reflects the input resistance of the cell and was chosen such that the map model matches a corresponding equivalent Hodgkin–Huxley model (Traub and Miles 1991), see Figure 1.7. The details of the corresponding model can be found in (Nowotny et al. 2005).

The 1000-fold speedup of the map model over a conventional Hodgkin-Huxley model allowed us to simulate learning in a system of approximately the size of the Drosophila olfactory pathway. The model system is illustrated in Figure 1.8. The

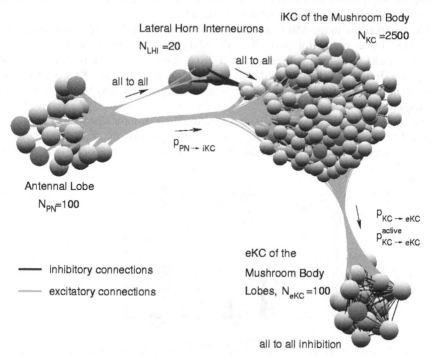

Fig. 1.8. More realistic model of the olfactory system of insects. The balls represent map-based neurons in a 10:1 ratio (i.e. each ball represents 10 neurons). The eKCs are connected all-to-all with mutual inhibition. The additional pathway involving lateral horn interneurons provides global inhibition on the KCs. (Modified from Nowotny et al. 2005).

20   T. Nowotny

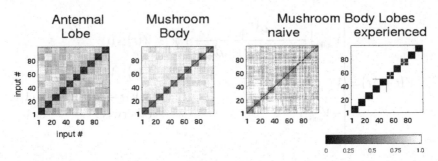

Fig. 1.9. Pairwise distances of activity patterns in the different stages (distances normalized between 0 and 1). Here, the system experiences 10 input classes of 10 patterns each. Patterns are sorted by classes, i.e., input 1 to 10 is class 1 etc. The distances are amplified in the projection from the AL to the MB. In the naïve MB lobes the structure of the activity patterns is almost lost while after sufficient experience classes are represented with maximal inter- and minimal intra-distance. The same result holds for up to 100 classes of 10 inputs each (not shown). (Modified from (Nowotny et al. 2005).

additional pathway through lateral horn interneurons provides a global inhibitory signal to the MB that is approximately proportional to the total activity in the AL. This inhibition implements a simple form of gain control and avoids the fat tails of the probability distribution of the number of active KCs.

The competition in the MB lobes is in this more detailed, time-resolved model implemented by a straightforward all-to-all inhibitory coupling between the eKCs. Whichever eKCs fire first will inhibit other eKCs and prevent them from firing at the same time. It turns out that this simple lateral interaction between eKCs is enough to implement the winner-take-all situation we previously imposed by hand and to shape the output code.

We again challenge the system with a structured set of inputs generated as before. Figure 1.9 illustrates our findings for a set of 10 input classes of 10 input patterns each. The grayscale maps in the figure show the pairwise distance of activity patterns

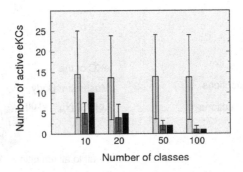

Fig. 1.10. Average number of eKCs representing each class in the naïve system (light gray), after experience (dark gray) and in an optimal disjoint representation (black). The experienced system almost reaches the optimal usage of neurons. (Modified from (Nowotny et al. 2005))

in the different parts of the system. The structure of the input set becomes apparent by the dark squares of small distance of inputs of the same class and lighter colors for the distance between patterns of different classes. All these distances are augmented by the projection to the MB manifested by the lighter colors. Before the system experiences any inputs, we call it the naïve system, the structure of the input set is almost lost in the MB lobes. After some experience, however, the distances within classes are minimal and those between classes are maximal (distances were normalized by the number of active neurons such that the maximum possible distance is 1). Note, that this process of "experience" only entails presentation of odor input patterns in the AL and no additional process of supervision or interference with the output patterns of any kind. The system self-organizes into a representation of identical output patterns for the patterns within each class and maximal difference of output patterns from different classes. Closer inspection reveals that the maximization of the difference between inter- and intra-distance occurs through completely disjoint representations, i.e., the exact same set of eKCs responds to every pattern of a given class and to no other pattern.

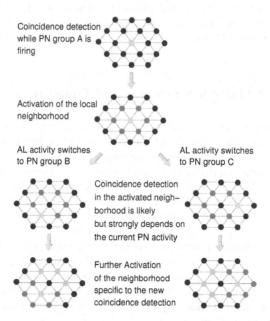

Fig. 1.11. Illustration of the disambiguation of temporal sequences of activity patterns. If a neuron is active (light grey circle) in one pattern the local excitatory connections activate the neighbors of the active neuron (dark gray circles). Activity in these neighborhoods during the next LFP cycle is now more probable than in other KCs. Which of the neighbors may fire a spike, however, still depends on the PN input during the next cycle. It might be a different neuron for a PN activity pattern B (left side) than for another pattern C (right side). In this way local sequences of active KC are formed which depend on the identity of active PNs as well as on the temporal order of their activity due to the activated neighborhoods. (Modified from (Nowotny et al. 2003).

The same experiment with 20, 50, and 100 input classes revealed similar results. For 100 input patterns the first failures appeared in the form of input patterns that did not elicit any response. This is not surprising as a set of 100 input classes represented with disjoint representations of active eKCs needs 100 output neurons from which each would represent one input class. As the number of eKCs was only 100 in total, 100 input classes is clearly the limit to which the system can be expected to perform.

The analysis of the representation of input classes by eKC firing patterns reveals another interesting effect. During "experience" the system is confronted with randomly chosen inputs from the input set. No direct information about the structure of this set is available to it. Nevertheless it manages to identify this structure and organize its output accordingly. As shown in Figure 1.10, the system uses about as many neurons as it can for each odor class while keeping the disjoint representation. It does not use more (not so surprising given the competition between neurons) but also not less. Just looking at the number of active eKCs for each input pattern one can tell the number of classes in the input set.

In summary, the more detailed models reveal how the nervous system of the locust may implement the elements of the odor classification scheme with random connections using simple elements like mutual all-to-all inhibition and Hebbian learning. We have also seen that the implementation of classification with these simple ingredients automatically solves the additional task of detecting the cluster structure of the input pattern set.

1.5   The Role of Multiple Snapshots and Time Integration

It has been shown repeatedly by careful analysis of the activity in the AL that the trajectory of AL activity (interpreted as the trajectory of a high-dimensional vector in an appropriate vector space spanned by the activity of single PNs) over time contains more salient information on the identity of an odour than a single snapshot (a point in this high-dimensional AL activity space) at any given time. Earlier we have argued that this activity trajectory is sliced into snapshots of 50 ms in the activity of KCs in the MB due to feedforward inhibition synchronized to the LFP in the AL/ MB calyx and the coincidence-detector property of the KCs themselves. Are we, therefore, to assume that the additional information that lies in the time evolution of the activity pattern in the AL is lost in the further processing and the insect does not make use of it? One solution to this problem could be time integration of several KC activity patterns in the eKCs. That, however, would still lead to some information loss. The sequence of patterns A, B, C and any permutation of it would look identical in the time-integrated activity patterns.

In a recent paper (Nowotny et al. 2003) we suggested a solution to this paradox based on the observation of the existence of lateral excitation between the axons of KCs (Leitch and Laurent 1996). We hypothesized that this lateral interaction could disambiguate the sequences A, B, C from its permutations by the mechanism illustrated in Figure 1.11.

To test this idea of "symmetry breaking" with lateral excitation, we have built a full size model of the locust olfactory system using Hodgkin-Huxley type model neurons for PNs, leaky integrate-and-fire neurons for KCs and a single Hodgkin-Huxley neuron to represent the global inhibition from lateral horn interneurons onto the KCs. The KCs

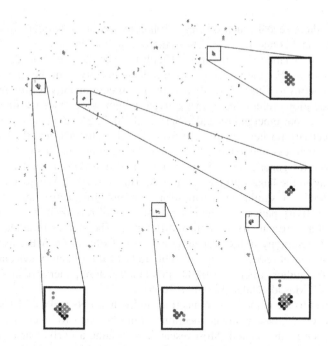

Fig. 1.12. Differences in the integrated activity of KCs. Black circles symbolize KCs that are more active in response to an activity pattern sequence A, C, B than to A, B, C, light grey circles symbolize KCs which activity is the same for both and dark grey circles are KCs which respond more to A, B, C. (Modified from (Nowotny et al. 2003)).

were connected by slow excitatory connections with the topology of a hexagonal lattice according to the earlier experimental findings (Leitch and Laurent 1996). When confronted with a PN activity pattern sequence A, B, C and its permutation A, C, B the integrated response of the KCs over these three LFP cycles is markedly different (Figure 1.12). This demonstrates that the lateral connections can help downstream integrators to distinguish the different sequences and provides a hypothetical function for the existing local, excitatory axo-axonal connections in the MB of locust.

In summary, the segmentation of AL activity into time discrete "snapshots" of activity in the KCs need not mean that the temporal dynamics in the AL does not matter or is not being used downstream. One could easily envision a system where single snapshots were used for rapid odour recognition and integrated activity for fine odour discrimination. The latter then would benefit from the additional disambiguation mechanism described in this section.

1.6 Discussion

Experimental Data on Connectivity

The experimental assessment of the connectivity between and within neuropils is a notoriously hard problem. Imaging studies with back-filled neurons and genetically expressed markers reveal much of the morphology but are often hard to interpret in terms of functional connections. Electron microscopy, on the other hand, allows identifying potential synaptic sites but it is difficult to trace the neural dendrites or axons of specific neurons. For the olfactory system, the connectivity of the first stages from ORNs to glomeruli to PNs is broadly accepted to predominantly follow the one OR-one ORN- one glomerulus- one PN rule, even though some evidence for more complicated schemes has emerged recently (Sato et al. 2007; Bhalerao et al. 2003).

The connectivity further downstream from the AL to the MB and the lateral protocerebrum is much less understood. It was mainly examined in *Drosophila* due to the available powerful genetic tools. Some studies find a striking reproducibility of axonal branching in the lateral protocerebrum (Marin et al. 2002; Wong et al. 2002; Tanaka et al. 2004). A similar observation of reproducibility across animals has been made on the activity patterns in the olfactory cortex of rats (Zou et al. 2005). In the MBs, the picture presents itself slightly differently. The studies in adult *Drosophila* find a lack of stereotypy in the branching patterns of PNs in the MB calyx (Wong et al. 2002; Marin et al. 2002) and in a recent study of the olfactory system in *Drosophila* larvae the authors refer to outright "potential random patterns of connectivity in the MB calyx" (Masuda-Nakagawa et al. 2005).

With respect to the number of connections individual PNs make on KCs in the MB calyx and, reversely, how many connections from PNs each KC receives, the experimental evidence is also mixed. Most researchers assume a fairly sparse connectivity based on the observation of just a few boutons of each PN in the MB calyx (Marin et al. 2002; Wong et al. 2002). These may, however, "form divergent synapses upon multitudinous surrounding Kenyon cell dendrites" (Yasuyama et al. 2002), such that the observation of on average 3-10 large boutons on each PN axon in the MB calyx of *Drosophila* may not imply only on the order of 10 post-synaptic KCs. A recent work using pairwise electrical recordings of PNs and KCs in the locust has suggested a more drastically different connectivity degree of approximately 50% of all-to-all connections (Jortner et al. 2007) which is hard to reconcile with the imaging results.

For models of the olfactory system the consequence is clear. One will have to rely on scanning wide ranges of possible parameter values until a clearer image of the experimental facts emerges.

Other Models

While many models have focused on the experimentally better described and more accessible AL (Getz and Lutz 1999; Av-Ron and Vibert 1996; Christensen et al. 2001; Linster et al. 1993; Linster and Smith 1997; Linster and Cleland 2001; Galan et al. 2004), see also (Clelland and Linster 2005; Laurent 2002) for recent reviews, some theories including higher brain centers have been proposed as well (Li and Hertz 2000; Hopfield 1991; 1999; Brody and Hopfield 2003; Schinor and Schneider 2001). The necessary experiments to validate any of these models of higher olfactory processing, including our own, are, however, still in development. In the discussion of our ideas of information processing based on random connectivities we would like to take the point of view that even if the olfactory system turns out not to be based on such a paradigm, we still may find inspiration for uses in novel technical applications.

Implications for Technical Applications

Recently, artificial general gas sensing devices have been built and equipped with bio-inspired odorant processing solutions (White et al. 1998; White and Kauer 1999; Fu et al. 2005). The strategy of nervous systems may, however, be quite different from technical solutions or the traditional artificial neural networks. Growing a large number of identical, simple neurons may be comparatively inexpensive and a random connectivity comes almost for free for biological systems. In this chapter I have presented evidence of recent years that systems making use of large numbers and random connectivities can be successful in tasks like pattern classification. Such neural networks with random connections may constitute a new paradigm of "sloppy engineering" in which the exact connections between elements do not matter.

The advantages are obvious, at least from a biological point of view. It is easier to grow such a system as no specific genetic encoding of connectivity and no complex axon targeting schemes are necessary. Furthermore, the system is inherently robust. Because the connections are randomly chosen in the first place, the destruction of some connections or the introduction of a few additional ones, will not compromise the overall functionality. The same applies to cell death or cell replacement.

How practical is such an approach for technological solutions? Clearly, as a solution for a CPU-based system, like a desktop computer, the approach is not very efficient. All neurons would have to be evaluated in sequence taking long computation times. In the framework of FPGAs (field programmable gate arrays), the sequential evaluation problem is absent as one can design massively parallel hardware, computing the state of all the neurons at the same time, similar to biological systems. However, random connections still do not seem natural in this framework as the designer would have to select those once and then implement them on the chip. Doing so, one could as well implement some specific, optimized connectivity and save some resources in terms of the number of connections and neurons. A platform suitable for this new "random computing" approach has yet to be designed. One would envision a self-organized system of many simple and cheap computing elements like the neurons this approach was inspired by.

Throughout the different analyses presented here, the fact that the considered activity patterns encode odors has not been used. Consequently the results should apply to general pattern classification problems as well. Our work in progress is demonstrating that a random classification machine as analyzed here can also be used to classify handwritten digits from the MNIST database quite successfully (Huerta and Nowotny 2007). Furthermore, no major modifications or additions to the main principle seem to be necessary to perform this new task.

1.7 Appendix A: Probability Distribution for the Number of Active KC

The probability for having n_{KC} out of N_{KC} active KCs for a given number of n_{PN} active PNs is given by (we suppress any reference to the time steps for clarity):

$$P(n_{KC} = k) = \sum_{l=0}^{N_{PN}} \binom{N_{PN}}{l} p_{PN}{}^l (1 - p_{PN})^{N_{PN} - l} P(n_{KC} = k \mid n_{PN} = l), \qquad (1.A1)$$

where the conditional probability is given by

$$P(n_{KC} = k \mid n_{PN} = l) = \binom{N_{KC}}{k} p_{KC}(l)^k (1 - p_{KC}(l))^{N_{KC} - k}. \qquad (1.A2)$$

Here, $p_{KC}(l)$ denotes the conditional probability for a single KC to fire given l PNs are active,

$$p_{KC}(l) = \sum_{j=\theta}^{l} \binom{l}{j} p_{PN \to KC}{}^j (1 - p_{PN \to KC})^{l - j}. \qquad (1.A3)$$

For a more explicit explanation see (Nowotny and Huerta 2003).

1.8 Appendix B: Collision Probability

Following the lead of (Garcia-Sanchez and Huerta 2003) we can write the probability that two activity patterns in the MB are identical as

$$P(\bar{y} = \bar{y}') = \left\langle \prod_{i=1}^{N_{KC}} (y_i y_i' + (1 - y_i)(1 - y_i')) \right\rangle_{\bar{x}, \bar{x}', C}, \qquad (1.B1)$$

where the triangular brackets denote the expectation value with respect to the product measure for inputs $\bar{x}$, $\bar{x}'$ and connectivity C.

The connectivity to each KC is chosen independently and clearly does not depend on the inputs present at any time. We can, therefore, pull the expectation over C inside, obtaining

$$P(\bar{y} = \bar{y}') = \left\langle \prod_{i=1}^{N_{KC}} \langle y_i y_i' + (1 - y_i)(1 - y_i') \rangle_C \right\rangle_{\bar{x}, \bar{x}'} \qquad (1.B2)$$

with

$$\prod_{i=1}^{N_{KC}} \langle y_i y_i' + (1 - y_i)(1 - y_i') \rangle_C = \left(\sum_{\{\bar{c}_1\}} (y_1 y_1' + (1 - y_1)(1 - y_1')) P(\bar{c}_1) \right)^{N_{KC}} \qquad (1.B3)$$

as all $\bar{c}_i$ (the columns of the connectivity matrix C) are independent and identically distributed (iid). Furthermore, $y_1 y_1' = 1$ only if y_1 is connected to θ or more active x_i and y_1' to θ or more active x_i'. This depends solely on the number of ones in $\bar{x}$

and in $\vec{x}'$, and how many of those are the same PNs. We denote the number of 1s in $\bar{x}$ but not $\bar{x}'$ by k, of 1s in $\bar{x}'$ but not in $\bar{x}$ by k' and the number of 1s in both by k''. Then

$$\sum_{\{\bar{c}_1\}} y_1 y_1' P(\bar{c}_1) = \sum_{j''=0}^{k''} \sum_{j'=\theta-j''}^{k'} \sum_{j=\theta-j'}^{k} b_{k'',p_c}(j'') b_{k',p_c}(j') b_{k,p_c}(j). \tag{1.B4}$$

Where $b_{k,p_c}(j) = \binom{k}{j} p_c^{\,j} (1-p_c)^{k-j}$ denotes the binomial distribution with pa-

rameters k and p_c and $p_c \equiv p_{\text{PN}\to\text{KC}}$ for the sake of brevity throughout this Appendix. The sums can be moved inside such that

$$\sum_{\{\bar{c}_1\}} y_1 y_1' P(\bar{c}_1) = \sum_{j''=0}^{k''} b_{k'',p_c}(j'') B_{k',p_c,\theta}(j'') B_{k,p_c,\theta}(j'') \tag{1.B5}$$

where $B_{k,p_c,\theta}(j'') = \sum_{j=\theta-j''}^{k} b_{k,p_c}(j)$. Similarly we obtain

$$\sum_{\{\bar{c}_1\}} (1-y_1)(1-y_1') P(\bar{c}_1) = \sum_{j''=0}^{k''} b_{k'',p_c}(j'')\left(1-B_{k',p_c,\theta}(j'')\right)\left(1-B_{k,p_c,\theta}(j'')\right). \tag{1.B6}$$

Denoting the sum of (B5) and (B6) as $A(k'',k',k)$ we obtain

$$P(\bar{y} = \bar{y}') = \langle A(k'',k',k) \rangle_{\bar{x},\bar{x}'} = \sum_{k'',k',k} A(k'',k',k) P(k'',k',k). \tag{1.B7}$$

The probability for triples k'',k',k then is

$$P(k'',k',k) = \binom{N_{\text{PN}}}{k''\,k'\,k} p_{\text{PN}}^{\,2k''} \left(p_{\text{PN}}(1-p_{\text{PN}})\right)^{k'+k} (1-p_{\text{PN}})^{2(N_{\text{PN}}-k-k'-k'')} \tag{1.B8}$$

where $\binom{N_{\text{PN}}}{k''\,k'\,k} = \dfrac{N_{\text{PN}}!}{k''!\,k'!\,k!\,(N_{\text{PN}}-k-k'-k'')!}$ is the multinomial coefficient.

The numerical evaluation of this result is rather computationally expensive but the simplification for (almost) independent y_i used in (Garcia-Sanchez and Huerta 2003) breaks down for larger p_c.

To be exact, we are not interested in the probability of collision per se but in the probability of a collision given the inputs $\bar{x}$ and $\bar{x}'$ do not collide, i.e.

$$P(\bar{y} = \bar{y}' | \bar{x} \neq \bar{x}') = P(\{\bar{y} = \bar{y}'\} \cap \{\bar{x} \neq \bar{x}'\}) \Big/ P(\bar{x} \neq \bar{x}'). \tag{1.B9}$$

Clearly, because of disjointness,

$$P(\{\bar{y} = \bar{y}'\} \cap \{\bar{x} \neq \bar{x}'\}) = P(\bar{y} = \bar{y}') - P(\{\bar{y} = \bar{y}'\} \cap \{\bar{x} = \bar{x}'\})$$
$$= P(\bar{y} = \bar{y}') - P(\bar{x} = \bar{x}'). \tag{1.B10}$$

and $P(\bar{x} = \bar{x}')$ is easily determined to be

$$P(\bar{x} = \bar{x}') = \left(p_{PN}^{2} + (1 - p_{PN})^{2}\right)^{N_{PN}}. \tag{1.B11}$$

In our main example $p_{PN} = 0.1$ leads to $P(\bar{x} = \bar{x}') = 3.46 \cdot 10^{-92}$ which can safely be neglected. In Figure 1.1B we thus used $P(\bar{y} = \bar{y}')$ directly.

1.9   Appendix C: Probability of Proper Classification for One Input

The activity patterns $\bar{y}_i$ are determined by the random choice of activity patterns $\bar{x}_i$ in the AL. We assume we have chosen the connectivity and firing threshold in the MB appropriately to ensure an (almost) one-to-one correspondence between $\bar{x}_i$ and $\bar{y}_i$. As the $\bar{x}_i$ are independently chosen the one-to-one relationship induces independence of the $\bar{y}_i$ such that the probability in (8) factorizes to

$$P_{classify}(n_{pattern}) = \sum_{l_{cherry}=0}^{N_{KC}} \prod_{i=1}^{n_{pattern}} P(z(\bar{y}_i) = 0 | l_{cherry}) P(n_{KC} = l_{cherry}) \tag{1.C1}$$

and because of symmetry the product is

$$\left(P(z(\bar{y}_1) = 0 | l_{cherry})\right)^{n_{pattern}}. \tag{1.C2}$$

After averaging over all PN-KC connectivities, all KC activity patterns with the same number n_{KC} of active KC are equally probable due to symmetry. Furthermore, the probability to have a pattern with given n_{KC} is given by the equations (1.A1)-(1.A3).

We have already conditioned by l_{cherry} and now also do so by the corresponding "length" for a new input $\bar{y}_1$, denoted as l_y, to use our knowledge on equipartition within the class of patterns with the same "length" l and our knowledge of $P(n_{KC} = l)$, i.e.,

$$P(z=0|l_{cherry}) = \sum_{l_y=0}^{N_{KC}} P(z=0|l_{cherry}, l_y) P(n_{KC}=l_y). \qquad (1.C3)$$

As we have seen, the connectivity vector $\bar{w}$ from KCs to the eKC is equal to the cherry KC activity pattern after sufficient experience. In order for the eKC not to fire to a non-cherry activity pattern, this pattern must share less than θ_{eKC} active KCs with the cherry pattern.

$$P(z=0|l_{cherry}, l_y) = \sum_{i=0}^{\theta_{eKC}-1} P(n_{overlap}=i|l_{cherry}, l_y). \qquad (1.C4)$$

Using the equipartition on the class of KC activity patterns with same n_{KC} patterns we can use direct combinatorics to obtain the conditional overlap probability for patterns of "length" l_{cherry} and l_y,

$$P(n_{overlap}=i|l_{cherry}, l_y) = \frac{\binom{l_{cherry}}{i}\binom{N_{KC}-l_{cherry}}{l_y-i}}{\binom{N_{KC}}{l_y}} \qquad (1.C5)$$

Combining the expressions (1.C2)-(1.C5) gives the full expression for $P(z=0)$ and was used to generate the one-odor classification results in Figure 1.3.

References

Av-Ron, E., Vibert, J.F.: A model for temporal and intensity coding in insect olfaction by a network of inhibitory neurons. Biosystems 39, 241–250 (1996)

Bhalerao, S., Sen, A., Stocker, R., Rodrigues, V.: Olfactory neurons expressing identified receptor genes project to subsets of glomeruli within the antennal lobe of Drosophila melanogaster. J. Neurobiol. 54, 577–592 (2003)

Brody, C.D., Hopfield, J.J.: Simple networks for spike-timing-based computation, with application to olfactory processing. Neuron. 37, 843–852 (2003)

Cazelles, B., Courbage, M., Rabinovich, M.I.: Anti-phase regularization of coupled chaotic maps modeling bursting neurons. Europhys. Lett. 56, 504–509 (2001)

Christensen, T.A., D'Alessandro, D., Lega, J., Hildebrand, J.G.: Morphometric modeling of olfactory circuits in the insect antennal lobe: I. simulations of spiking local interneurons. Biosystems 61, 143–153 (2001)

Cleland, T.A., Linster, C.: Computation in the olfactory system. Chem. Senses 30, 801–813 (2005)

Cortes, C., Vapnik, V.: Support vector networks. Mach. Learn. 20, 273–297 (1995)

Cover, T.: Geometric and statistical properties of systems of linear inequalities with applications in pattern recognition. IEEE T. Electron. Comput. 14, 326 (1965)

Ernst, K.D., Boeckh, J., Boeckh, V.: A neuroanatomical study of the organization of the central antennal pathways in insects. Cell Tissue Res. 329, 143–162 (1977)

Farivar, S.S.: Cytoarchitecture of the Locust Olfactory System. Dissertation, California Institute of Technology (2005)

Feinstein, P., Bozza, T., Rodriguez, I., Vassalli, A., Mombaerts, P.: Axon guidance of mouse olfactory sensory neurons by odorant receptors and the beta2 adrenergic. Cell 117, 833–846 (2004)

Feinstein, P., Mombaerts, P.: A contextual model for axonal sorting into glomeruli in the mouse olfactory system. Cell 117, 817–831 (2004)

Friedrich, R.W., Laurent, G.: Dynamic optimization of odor representations by slow temporal patterning of mitral cell activity. Science 291, 889–894 (2001)

Fu, J., Yang, X., Yang, X., Li, G., Freeman, W.: Application of biologically modeled chaotic neural network to pattern recognition in artificial olfaction. In: Conf. Proc. IEEE Eng. Med. Biol. Soc., vol. 5, pp. 4666–4669 (2005)

Gao, Q., Yuan, B., Chess, A.: Convergent projections of Drosophila olfactory neurons to specific glomeruli in the antennal lobe. Nat. Neurosci. 3, 780–785 (2000)

Galan, R.F., Sachse, S., Galizia, C.G., Herz, A.V.M.: Odor-driven attractor dynamics in the antennal lobe allow for simple and rapid olfactory pattern classification. Neural Comput. 16, 999–1012 (2004)

García-Sanchez, M., Huerta, R.: Design Parameters of the Fan-Out Phase of Sensory Systems. J. Comput. Neurosci. 15, 5–17 (2003)

Getz, W.M., Lutz, A.: A neural network model of general olfactory coding in the insect antennal lobe. Chem. Senses 24, 351–372 (1999)

Gill, D.S., Pearce, T.C.: Wiring the olfactory bulb - activity-dependent models of axonal targeting in the developing olfactory pathway. Rev. Neurosci. 14, 63–72 (2003)

Hansson, B.S., Ochieng, S.A., Grosmaitre, X., Anton, S., Njagi, P.G.N.: Physiological responses and central nervous projections of antennal olfactory receptor neurons in the adult desert locust, Schistocerca gregaria (Orthoptera: Acrididae). J. Comp. Physiol. A 179, 157–167 (1996)

Hopfield, J.J.: Olfactory computation and object perception. P. Natl. Acad. Sci. USA 88(15), 6462–6466 (1991)

Hopfield, J.J.: Odor space and olfactory processing: collective algorithms and neural implementation. P. Natl. Acad. Sci. USA 96, 12506–12511 (1999)

Huerta, R., Nowotny, T., Garcia-Sanchez, M., Abarbanel, H.D.I., Rabinovich, M.I.: Learning classification in the olfactory system of insects. Neural Comput. 16, 1601–1640 (2004)

Huerta, R., Nowotny, T.: Classifying handwritten numbers with the "insect mushroom bodies (in preparation) (2007)

Jortner, R.A., Farivar, S.S., Laurent, G.: A Simple Connectivity Scheme for Sparse Coding in an Olfactory System. J. Neurosci. 27, 1659–1669 (2007)

Laurent, G., Wehr, M., Davidowitz, H.: Temporal representations of odors in an olfactory network. J. Neurosci. 16, 3837–3847 (1996)

Laurent, G.: A systems perspective on early olfactory coding. Science 286, 723–728 (1999)

Laurent, G., Stopfer, M., Friedrich, R.W., Rabinovich, M.I., Abarbanel, H.D.I.: Odor encoding as an active, dynamical process: Experiments, computation, and theory. Annu. Rev. Neurosci. 24, 263–297 (2001)

Laurent, G.: Olfactory network dynamics and the coding of multidimensional signals. Nat. Rev. Neurosci. 3, 884–895 (2002)

Leitch, B., Laurent, G.: GABAergic synapses in the antennal lobe and mushroom body of the locust olfactory system. J. Comp. Neurol. 372, 487–514 (1996)

Li, Z., Hertz, J.: Odour recognition and segmentation by a model olfactory bulb and cortex. Network 11, 83–102 (2000)

Linster, C., Cleland, T.A.: How spike synchronization among olfactory neurons can contribute to sensory discrimination. J. Comput. Neurosci. 10, 187–193 (2001)

Linster, C., Masson, C., Kerszberg, M., Personnaz, L., Dreyfus, G.: Computational diversity in a formal model of the insect olfactory macroglomerulus. Neural Comput. 5, 228–241 (1993)

Linster, C., Smith, B.H.: A computational model of the response of honey bee antennal lobe circuitry to odor mixtures: Overshadowing, blocking and unblocking can arise from lateral inhibition. Behav. Brain Res. 87, 1–14 (1997)

Marin, E.C., Jefferis, G.S., Komiyama, T., Zhu, H., Luo, L.: Representation of the glomerular olfactory map in the Drosophila brain. Cell 109, 243–255 (2002)

Masuda-Nakagawa, L.M., Tanaka, N.K., O'Kane, C.J.: Stereotypic and random patterns of connectivity in the larval mushroom body calyx of Drosophila. Proc. Natl. Acad. Sci. USA 102, 19027–19032 (2005)

McCulloch, W.S., Pitts, W.: Logical Calculus of Ideas Immanent in Nervous Activity. B. Math. Biophys. 5, 115–133 (1943)

Nowotny, T., Huerta, R.: Explaining Synchrony in feed-forward networks: Are McCulloch-Pitts neurons good enough? Biol. Cybern. 89, 237–241 (2003)

Nowotny, T., Rabinovich, M.I., Huerta, R., Abarbanel, H.D.I.: Decoding temporal information through slow lateral excitation in the olfactory system of insects. J. Comput. Neurosci. 15, 271–281 (2003)

Nowotny, T., Huerta, R., Abarbanel, H.D.I., Rabinovich, M.I.: Self-organization in the olfactory system: one shot odor recognition in insects. Biol. Cybern. 93, 436–446 (2005)

Perez-Orive, J., Mazor, O., Turner, G.C., Cassenaer, S., Wilson, R.I., Laurent, G.: Oscillations and sparsening of odor representations in the mushroom body. Science 297(5580), 359–365 (2002)

Rabinovich, M.I., Volkovskii, A., Lecanda, P., Huerta, R., Abarbanel, H.D.I., Laurent, G.: Dynamical Encoding by Networks of Competing Neuron Groups: Winnerless Competition. Phys. Rev. Lett. 87, 068102 (2001)

Rulkov, N.F.: Modeling of spiking-bursting behavior using two-dimensional map. Phys. Rev. E 65, 041922 (2002)

Sato, Y., Miyasaka, N., Yoshihara, Y.: Hierarchical regulation of odorant receptor gene choice and subsequent axonal projection of olfactory sensory neurons in zebra fish. J. Neurosci. 27, 1606–1615 (2007)

Schinor, N., Schneider, F.W.: A small neural net simulates coherence and short-term memory in an insect olfactory system. Phys. Chem. Chem. Phys. 3, 4060–4071 (2001)

Scott, K., Brady Jr., R., Cravchik, A., Morozov, P., Rzhetsky, A., Zuker, C., Axel, R.: A chemosensory gene family encoding candidate gustatory and olfactory receptors in Drosophila. Cell 104, 661–673 (2001)

Tanaka, N.K., Awasaki, T., Shimada, T., Ito, K.: Integration of chemosensory pathways in the Drosophila second-order olfactory centers. Curr. Biol. 14, 449–457 (2004)

Tozaki, H., Tanaka, S., Hirata, T.: Theoretical consideration of olfactory axon projection with an activity-dependent neural network model. Mol. Cell Neurosci. 26, 503–517 (2004)

Traub, R.D., Miles, R.: Neural networks of the hippocampus. Cambridge University Press, New York (1991)

Vosshall, L.B., Wong, A.M., Axel, R.: An olfactory sensory map in the fly brain. Cell 102, 147–159 (2000)

Wang, J.W., Wong, A.M., Flores, J., Vosshall, L.B., Axel, R.: Two-photon calcium imaging reveals an odor-evoked map of activity in the fly brain. Cell 112, 271–282 (2003)

Wang, Y., Wright, N.J., Guo, H., Xie, Z., Svoboda, K., Malinow, R., Smith, D.P., Zhong, Y.: Genetic manipulation of the odor-evoked distributed neural activity in the Drosophila mushroom body. Neuron. 29, 267–276 (2001)

Wehr, M., Laurent, G.: Odour encoding by temporal sequences of firing in oscillating neural assemblies. Nature 384(6605), 162–166 (1996)

White, J., Dickinson, T.A., Walt, D.R., Kauer, J.S.: An olfactory neuronal network for vapor recognition in an artificial nose. Biol. Cyber. 78, 245–251 (1998)

White, J., Kauer, J.S.: Odor recognition in an artificial nose by spatio-temporal processing using an olfactory neuronal network. Neurocomputing 26-27, 919–924 (1999)

Wong, A.M., Wang, J.W., Axel, R.: Spatial representation of the glomerular map in the Drosophila protocerebrum. Cell 109, 229–241 (2002)

Wüstenberg, D.G., Boytcheva, M., Grünewald, B., Byrne, J.H., Menzel, R., Baxter, D.A.: Current- and voltage-clamp recordings and computer simulations of Kenyon cells in the honeybee. J. Neurophysiol. 92, 2589–2603 (2004)

Yasuyama, K., Meinertzhagen, I.A., Schurmann, F.W.: Synaptic organization of the mushroom body calyx in Drosophila melanogaster. J. Comp. Neurol. 445, 211–226 (2002)

Zou, Z., Li, F., Buck, L.B.: Odor maps in the olfactory cortex. Proc. Natl. Acad. Sci. USA 102, 7724–7729 (2005)

2

From ANN to Biomimetic Information Processing

Anders Lansner[1,2], Simon Benjaminsson[1], and Christopher Johansson[1]

[1] Royal Institute of Technology (KTH), Dept Computational Biology
[2] Stockholm University, Dept Numerical Analysis and Computer Science,
AlbaNova University Center SE - 106 91 Stockholm
{ala,simonbe,cjo}@csc.kth.se

Abstract. Artificial neural networks (ANN) are useful components in today's data analysis toolbox. They were initially inspired by the brain but are today accepted to be quite different from it. ANN typically lack scalability and mostly rely on supervised learning, both of which are biologically implausible features. Here we describe and evaluate a novel cortex-inspired hybrid algorithm. It is found to perform on par with a Support Vector Machine (SVM) in classification of activation patterns from the rat olfactory bulb. On-line unsupervised learning is shown to provide significant tolerance to sensor drift, an important property of algorithms used to analyze chemo-sensor data. Scalability of the approach is illustrated on the MNIST dataset of handwritten digits.

2.1 Introduction

Artificial neural networks and related learning based techniques add important functionality to today's signal processing and data analysis toolboxes. In particular, such methods excel in supervised learning and e.g. SVM challenges human performance in specific domains like recognition of isolated handwritten digits. These methods were initially inspired and motivated by analogies with the brain, but today this connection is rarely emphasized. On the contrary, ANN:s are in many aspects different from biology, for instance, by their lack of scalability to brain-sized networks, their focus on deterministic computing, and on supervised learning based on the availability of labelled training examples. All of these features are markedly non-biological.

Current knowledge about the brain suggests that its architecture is highly scalable and run on stochastic computing elements which employ Hebbian type correlation and reinforcement based learning rules rather than supervised ones. In fact, supervised error correction learning techniques are quite suspect from the point of view of neurobiology. Thorpe and Imbert reviewed the arguments some time ago but their remarks are still valid (Thorpe and Imbert 1989). Quinlan suggested that, in fact, the multi-layer perceptron is super-competent on many tasks compared to humans, which reduces its plausibility as models of the brain (Quinlan 1991).

Why should we be interested in neurobiology at all? Well, in important respects, our man-made methods and artefacts still lag far behind biological systems. The latter excel in real-time, real world perception and control, handling of input from high dimensional sensor arrays, as well as holistic pattern recognition including

A. Gutiérrez and S. Marco (Eds.): Biologically Inspired Signal Processing, SCI 188, pp. 33–43.
springerlink.com   © Springer-Verlag Berlin Heidelberg 2009

figure-ground separation and information fusion. They also demonstrate exceptional compactness, tolerance to hardware faults and low energy consumption. These are attractive properties also from a technological perspective.

With the increasing abundance of sensors and sensor arrays as well as massive amounts of data generated in many different applications of advanced information technology and autonomous systems there is an increasing technical interest in scalable and unsupervised approaches to learning-based data analysis and in robotics. Also, as new molecular scale computing hardware is developed, the interest in robust algorithms for stochastic computing will increase.

A serious complication is that we do not yet fully understand the computational and information processing principles underlying brain function. An increasingly important tool in brain science is quantitative modelling and numerical simulation. In the field of computational neuroscience models at different levels of biophysical detail

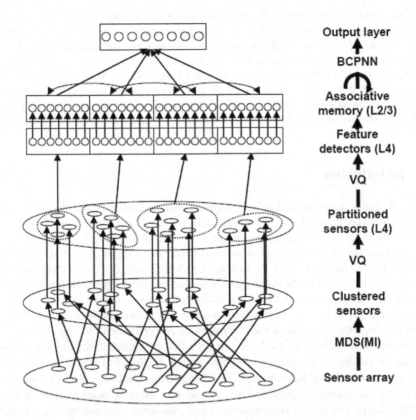

Fig. 2.1. Outline of the hybrid algorithm. The unstructured array of sensors is clustered using multi-dimensional scaling (MDS) with a mutual information (MI) based distance measure. Then Vector Quantization (VQ) is used to partition the sensor into correlated groups. Each such group provides input to one module of an associative memory layer. VQ is used again to provide each module unit with a specific receptive field, i.e. to become a feature detector. Finally, classification is done by means of BCPNN.

are developed and investigated to speed up the development of our understanding of how the brain works (De Schutter et al. 2005). In fact, the most reduced models in this field are formulated on a level of abstraction close to that of ANN, so called connectionist models. Such models can serve as the starting point for the design of brain-inspired computational structures, and this is the approach we have taken.

Our overall goal is the development of a generic cortex-inspired computational building block that allows for the design of modular and recursive hierarchical adaptive pattern processing structures useful in technical applications as those mentioned above. This development is in its early stages, and we report here the basic design and evaluation of a novel hybrid algorithm aimed for this purpose.

2.1.1 The Underlying Abstract Model of Cortex

We have previously developed and investigated biophysically detailed models of the associative memory function of neocortex based on experimental data (Lundqvist et al. 2006). Based on the knowledge gained we have formulated an abstract network model of cortical layers 2/3 that forms the core of our present approach (Lansner and Holst 1996; Sandberg et al. 2002; Johansson and Lansner 2006a). Layer 5 is also likely to be closely interacting with layers 2/3 and is not represented separately (Hirsch and Martinez 2006).

An important additional operation is the transformation from raw sensor data to the sparse and distributed representations employed in cortical layer 2/3. This transformation is started in the early sub-cortical sensory processing streams but is continued in the forward pathway of cortex that involves its layer 4 as a key component. In our abstract model we represent layer 4 separately as a layer that self-organizes a modular ("hypercolumnar") structure and also decorrelates the input forming specific receptive fields and response properties of units in this layer. The hypercolumnar structure is imposed on layer 2/3 when formed and the layer 4 units drive their companion units in layer 2/3 via specific one-to-one connections. In the simplest case, as in the simulations described in the following, there is a feedforward projection from layer 2/3 to some output layer. In general, this structure can be extended recursively with projections connecting layer 2/3 to a layer 4 in the next level in the hierarchy located in a different cortical area. Long-range recurrent connections may also form between hypercolumns within layer 2/3 at the same level, forming the basis for autoassociation.

2.2 Methods

The proposed algorithm for one module works in several stages (Figure 2.1). First a sensor clustering followed by a vector quantization step partitions the input space. Then each group is decorrelated and sparsified in a feature extraction step, again using vector quantization. Finally the data is fed into an associative memory which is used in a feed-forward classification setting. Each step is explained in detail below.

2.2.1 Partitioning of Input Space

We consider the case of sensors with discrete coded values or value intervals. For sensor X and Y, the general dependence is calculated by the mutual information

$$I(X,Y) = \sum_{i \in Y} \sum_{j \in X} p_{ij} \log \frac{p_{ij}}{p_i p_j} \tag{2.1}$$

Here, i and j are the indexes for the units in each hypercolumn and the probabilities are estimated as

$$p_i = \frac{1}{P} \sum_{\mu=1}^{P} \xi_i^{\mu} \tag{2.2}$$

$$p_{ij} = \frac{1}{P} \sum_{\mu=1}^{P} \xi_i^{\mu} \xi_j^{\mu} \tag{2.3}$$

Where P is the number of input patterns and is the unit value at position i for input pattern μ. In case of continuous variables the values in this step needs to be interval coded.

The mutual information is transformed into a distance measure (Kraskov et al. 2005):

$$D(X,Y) = 1 - \frac{I(X,Y)}{J(X,Y)} \tag{2.4}$$

with the joint entropy calculated as

$$J(X,Y) = -\sum_{i \in Y} \sum_{j \in X} p_{ij} \log p_{ij} \tag{2.5}$$

From the full distance measure matrix we can create a multidimensional geometric map fulfilling the distance relations by employing classical multidimensional scaling (Young 1985). The number of dimensions in this map is specified to be as low as possible (without reducing the quality of the map too much) in order to reduce the computational needs in the following step. The number of partitions of the input space is manually specified and sets the number of code vectors in a vector quantization (VQ, see below) of the map produced by the multidimensional scaling. The VQ encoding process on each element in the map decides which group each sensor should belong to. The sensors with high general dependences (as determined by the mutual information) will in this way be grouped together.

2.2.2 Decorrelation and Sparsification

For each group, we perform VQ on the input from the subset sensors that belongs to that specific group, resulting in a decorrelated and sparsified code well suited for an associative memory system (Steinert et al. 2006). The VQ is performed by means of Competitive Selective Learning (CSL) (Ueda and Nakano 1994), but another VQ algorithm could have been used. As for standard competitive learning, CSL updates the weight from an input unit i to the output unit with highest activity for input pattern ξ^{μ} as

$$w_{ij} = w_{ij} + \epsilon(\xi_i^{\mu} - w_{ij}) \tag{2.6}$$

ϵ is the step length of change which decreases during learning. CSL also adds a selection mechanism which avoids local minima by reinitializing weight vectors.

2.2.3 Associative Memory

The resulting decorrelated and sparsified code ξ'^{μ} for input pattern μ is fed into a BCPNN (Bayesian Confidence Propagating Neural Network) with hypercolumns that uses a supervised correlation based Bayesian learning algorithm (Johansson and Lansner 2006). Here we are only interested in classification, so the input code from the intermediate layer is directly mapped to an output layer having a single hypercolumn, using a feed-forward pass where the weights are learned by the Hebbian-Bayesian learning. If we consider the output code from a group in the intermediate layer as a hypercolumn Qg, where each unit corresponds to a code vector from the VQ, and the classes as units in the output hypercolumn, the weight between presynaptic unit and postsynaptic unit is computed as

$$w_{ij} = \begin{cases} 0 & p'_i = 0 \vee p'_j = 0 \\ 1/P & p'_{ij} = 0 \\ \dfrac{p'_{ij}}{p'_i p'_j} & otherwise \end{cases} \qquad (2.7)$$

and p'_i and p'_{ij} are probabilities once again estimated according to Eqs. 2.2 and 2.3 above.

For each generated input pattern ξ'^{μ}. Each unit has a bias set to be

$$\beta_j = \begin{cases} \log 1/P^2 & p'_j = 0 \\ \log p'_j & otherwise \end{cases} \qquad (2.8)$$

When an incoming pattern is processed the activity in postsynaptic unit is calculated as

$$s_j = \log \beta_j + \sum_g \log \sum_{i \in Q_g} w_{ij} o_i \qquad (2.9)$$

Here we sum over all groups and all units in each group where is the activation value of unit i.

The final output is calculated by a softmax function, controlled by the gain parameter G, over all the units in the output layer:

$$o_j = \frac{e^{Gs_j}}{\sum_{k \in Q_o} e^{Gs_k}} \qquad (2.10)$$

In a classification task, the unit with the highest output is taken as the classification result.

2.2.4 Data Sets

In this study we used two different datasets, one of activation patterns from rat olfactory bulb and one of isolated handwritten digits.

Rat Olfactory Bulb Activation Patterns

The olfactory bulb activation data of Leon and Johnson was used as one of the evaluation data sets (Leon and Johnson). We used a subset comprising 2-deoxyglucose (2-DG) imaged activation patterns from 94 different odour stimuli. These spatial activation patterns were clustered in 60 different local spatial clusters. The mean activity within each such cluster was transformed to the range [0,1], whereby 94 patterns with 60 components were obtained (Marco et al. 2006).

The classification task was to separate these compounds into their chemical classes, i.e. acids (24), aldehydes (19), alcohols (16), ketones (17), esters (6), hydrocarbons (8), and misc (4). A random subset of 75% of these patterns was used for training and the rest comprised the validation set.

The robustness to sensor drift of the method under study was evaluated using a simple synthetic drift model. A gain for each of the 60 sensors was initiated to 1 after which the gain factor was subject for over 100 random-walk steps taken from a Gaussian distribution with = 0.01. In the on-line learning condition while testing drift robustness, the last unsupervised vector quantization step was run continuously.

MNIST Data

The MNIST data set consists of handwritten images, 28x28 pixels large with 256 gray levels (Figure 2.2). It has a training set of 60,000 samples and a test set of 10,000 samples. Specialized classifiers based on SVM have been reported to be more than 99% correct on the test set while a standard single layered network typically achieves 88% with no preprocessing (LeCun et al. 1998).

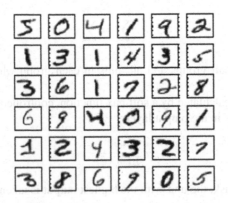

Fig. 2.2. Samples from the MNIST data set

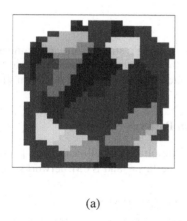   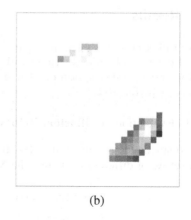

(a)   (b)

Fig. 2.3. Patches generated from the MNIST data by the MI + MDS + VQ + VQ steps of the hybrid algorithm. (a) The 12 different patches are colour coded. Note that some patches comprise more than one subfield. (b) Example of the specific receptive field of one of the 10 units in the patch marked with orange (with two subfields).

We can illustrate each step in the previous section by applying the proposed method to a real classification task.

In our example, the bit depth of all MNIST images is lowered by reducing the number of gray levels to eight. One input hypercolumn, corresponding to one image pixel, then can take on eight different values.

The general dependences between the image pixels are calculated by the mutual information. After multidimensional scaling the resulting matrix is grouped into P partitions by performing vector quantization. The result is a 28x28 map which shows how the pixels should be grouped, see Figure 2.3a for the case $P = 12$. Note that this is an entirely data driven approach that is independent of sensor modality. In the case of images, this step replaces the commonly used square tiling of the image. However, such tiling can only be applied when the correlation structure of the data is known beforehand to be two-dimensional.

We again perform vector quantization on each subset of sensors and form Q code vectors for each group. This gives us a total of $P \cdot Q$ units in the intermediate layer between the input and associative layer. Each code vector corresponds to a receptive field, an example of which is seen in Figure 2.3b, where we have backtracked the connections between a single code vector and the input sensors in a setting where $Q = 10$.

2.2.5 MLP/BP and SVM Software

The MLP/BP code used here to process the olfactory bulb activation patterns was from MATLAB® 7.3.0 NN-toolbox, using the scaled conjugate gradient learning rule with weight regularization. The SVM code used the osu-svm toolbox for MATLAB® (Ma et al. 2006). Parameters were in both cases selected to obtain best average performance on the validation set. Average and SEM of classification performance were calculated.

2.3 Results

This result section has three main parts, the first showing a straight-forward comparison of our novel hybrid algorithm with other methods, the second demonstrating the drift-tolerance of this algorithm relative to other methods, and the third demonstrating its scaling performance.

2.3.1 Evaluation on Olfactory Bulb Activation Patterns

We compared the results on the classification of the olfactory bulb activation pattern data set using different methods. The MLP/BP, one-layer, and SVM networks were

Table 2.1. Classification performance on validation set

METHOD	%correct (validation)
Onelayer	45 %
MLP/BP w reg	64%
SVM (Poly)	66%
SVM (RBF)	70%
VQ-BCPNN (1)	69%
VQ-BCPNN (7)	60%

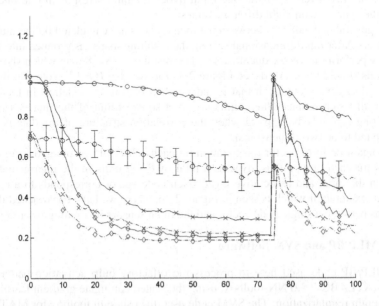

Fig. 2.4. Drift robustness of SVM, new method and new on-line learning method. Solid and dash-dotted lines represent performance on training and test sets respectively. Diamond, cross and circle refers to SVM, new method, and new on-line method respectively. Error bars are given only for performance of new on-line method on test data. At step 75 a complete recalibration is performed.

run as described in the methods section. The hybrid algorithm was run using two set-ups, one with just a single sensor partition and the other with the 60 sensors partitioned into seven groups. The total number of units in the BCPNN input layer was 70 in both cases. The results of this comparison are given in Table 2.1. The hybrid algorithm performs on par with SVM when only a single partition is used.

Drift tolerance was tested according to the description above using this data set and results are shown in Figure 2.4. As can be seen, the new algorithm with on-line learning has a much superior drift tolerance under these conditions.

2.3.2 Classification of MNIST Data

The algorithm was run on the entire MNIST data set. Of the 10,000 images in the test set, 95% were correctly classified (Johansson 2006). When only a feedforward configuration of BCPNN was used, with no intermediate layer generated by the hybrid algorithm, 84% of the images were correctly classified. Note that the learning in this case is not gradient descent but one-shot and correlation based.

Scaling performance of the algorithm and its dependence on the number of units in each hypercolumn is illustrated in Figure 2.5. As can be seen, the performance levels off at about 95% when there are more than one hundred units in each hypercolumn. Since there are eight hypercolumns, the total number of units in the internal layer is in this case up to one thousand.

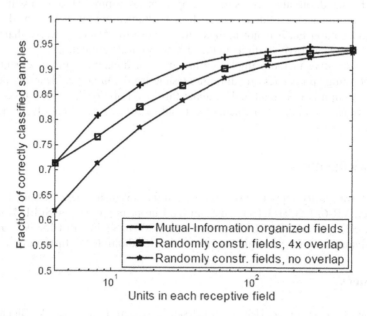

Fig. 2.5. Scaling performance of the new hybrid algorithm. Dependence of classification performance on the number of units in each hypercolumn (Johansson 2006).

2.4 Discussion and Conclusions

In this study of the performance of a novel hybrid algorithm for pattern processing we have proposed and described the different steps of the algorithm and evaluated its performance with regard to classification, drift robustness, and scaling. The algorithm is entirely data driven and does not make any assumptions on correlation structure, e.g. when processing image data.

When compared to an rbf-SVM approach on a small dataset of olfactory bulb activation patterns the new algorithm performed on par with SVM when no subdivision of the input space was done. With subdivision of the input space in seven disjoint groups, performance dropped significantly, from 70% to 60%. This suggests that the algorithm did in this case not find a set of seven independent groups of sensors. No method was able to reach beyond 70% which suggests that the problem is not separable, even non-linearily so. Comparison using more standard pattern classification benchmark datasets is ongoing.

In the test of robustness to sensor drift it was shown that when the unsupervised part of the algorithm was allowed to run in on-line training mode drift robustness much superior to SVM and the new algorithm with no on-line learning was demonstrated. This is a promising result, but further characterization of this property is required. Additional evaluation is currently ongoing on a real chemosensor dataset.

On the MNIST dataset the algorithm was able to reach 95% performance on the test set. This is not as good as a carefully designed SVM that reaches beyond 99%. On the other hand, our aim here is to develop a cortex-inspired algorithm with similar learning performance as a human being. It is not known how well humans do on the full MNIST dataset but it is not unlikely to be close to 95% (e.g. worse than SVM) given that many handwritten digits in this dataset are truly ambiguous.

Since associative memory implementing attractor dynamics, reinforcement learning and boosting approaches are all quite feasible from a biological learning perspective our ambition is to extend and evaluate this novel approach in such tasks and to focus on scalable parallel implementation to allow processing of data from arrays of millions of sensors.

Acknowledgements

This work was partly supported by grants from the Swedish Science Council (Vetenskapsrådet, VR-621-2004-3807) and from the European Union (GOSPEL NoE FP6-2004-IST-507610 and FACETS project, FP6-2004-IST-FETPI-015879), and the Swedish Foundation for Strategic Research (via Stockholm Brain Institute).

References

De Schutter, E., Ekeberg, Ö., Hellgren Kotaleski, J., Achard, P., Lansner, A.: Biophysically detailed modelling of microcircuits and beyond. Trends Neurosci. 28, 562–569 (2005)

Hirsch, J.A., Martinez, L.M.: Laminar processing in the visual cortical column. Curr. Opin. Neurobiol. 16, 377–384 (2006)

Johansson, C.: An attractor memory model of neocortex. School of Computer Science and Communication. Stockholm, Royal Institute of Technology, Sweden. PhD (2006)

Johansson, C., Lansner, A.: Attractor memory with self-organizing input. In: Ijspeert, A.J., Masuzawa, T., Kusumoto, S. (eds.) BioADIT 2006. LNCS, vol. 3853, pp. 265–280. Springer, Heidelberg (2006)

Johansson, C., Lansner, A.: A hierarchical brain-inspired computing system. In: International Symposium on Nonlinear Theory and its applications (NOLTA), Bologna, Italy, pp. 599–603 (2006b)

Kraskov, A., Stögbauer, H., Andrzejak, R.G., Grassberger, P.: Hierarchical clustering using mutual information. Europhys. Lett. 70(2), 278 (2005)

Lansner, A., Holst, A.: A higher order Bayesian neural network with spiking units. Int. J. Neural Systems 7(2), 115–128 (1996)

LeCun, Y.L., Bottou, L., Bengio, Y., Haffner, P.: Gradient-based learning applied to document recognition. Proc. IEEE 86(11), 2278–2324 (1998)

Leon, M., Johnson, B.A.: Glomerular activity response archive (2006), http://leonserver.bio.uci.edu

Lundqvist, M., Rehn, M., Djurfeldt, M., Lansner, A.: Attractor dynamics in a modular network model of the neocortex. Network: Computation in Neural Systems 17, 253–276 (2006)

Ma, J., Zhao, Y., Ahalt, S., Eads, D.: OSU-SVM for Matlab (2006), http://svm.sourceforge.net

Marco, S., Lansner, A., Gutierrez Galvez, A.: Exploratory analysis of the rat olfactory bulb activity. Abstract. ECRO 2006, Granada, Spain (2006)

Quinlan, P.: Connectionism and psychology. A psychological perspective on connectionist research. New York, Harvester, Whaetsheaf (1991)

Sandberg, A., Lansner, A., Petersson, K.-M., Ekeberg, Ö.: Bayesian attractor networks with incremental learning. Network: Computation in Neural Systems 13(2), 179–194 (2002)

Steinert, R., Rehn, M., Lansner, A.: Recognition of handwritten digits using sparse codes generated by local feature extraction methods. In: 14th European Symposium on Artificial Neural Networks (ESANN) 2006, Brugge, Belgium, pp. 161–166 (2006)

Thorpe, S., Imbert, M.: Biological constraints on connectionist modelling. In: Pfeiffer, R., Berlin, E. (eds.) Connectionism in Perspective. Springer, Berlin (1989)

Ueda, N., Nakano, R.: A new competitive learning approach based on an equidistortion principle for designing optimal vector quantizers. Neural Networks 7(8), 1211–1227 (1994)

Young, F.W.: Multidimensional scaling. Encyclopedia of Statistical Sciences 5, 649–659 (1985)

3

The Sensitivity of the Insect Nose: The Example of *Bombyx Mori*

Karl-Ernst Kaissling

Max-Planck-Institut fuer Verhaltensphysiologie. Seewiesen, 82319 Starnberg, Germany
Kaissling@orn.mpg.de

Abstract. The male silkmoth *Bombyx mori* responds behaviourally to bombykol concentrations in air of 3,000 molecules per ml presented at an air speed of 57 cm/s, i.e. the moth is almost as sensitive as a dog. The number of bombykol receptor neurons per antenna is 17,000, about 10,000-fold smaller than olfactory neurons found in dog noses. This high sensitivity is possible due to a very effective capture of odorant molecules and transport to the receptor neurons. The effectiveness of the insect antenna/nose has been determined by using radiolabeled bombykol, counting nerve impulses generated by the receptor neuron, and measuring the behavioural response of the male moth. At the behavioural threshold the neuronal signal/noise discrimination works at the theoretical limit.

3.1 Introduction

For a low olfactory threshold several sensory functions need to be optimized. Odour molecules have to be a) effectively caught by the antenna from the air space, and b) conducted with little loss to the olfactory receptor neurons. c) The odour stimulus has to be most sensitively transduced into nerve impulses, and d) the stimulus-induced impulse firing has to be distinguished from the background of spontaneous impulse discharge from the unstimulated receptor neurons. This paper reviews quantitative work on these items in the male moth of a species which is attracted (i.e. stimulated to walk upwind, Kaissling 1997) by a single chemical pheromone component, (E,Z)-10, 12-hexadecadienol (bombykol) released by the female moth (Butenandt et al. 1959).

3.2 Molecule Capture by the Antenna

To investigate the effectiveness of molecule capture by the antenna we used ^{3}H-labelled bombykol (Kasang 1968; Schneider et al 1968). With a high specific activity of 31.7 Ci/g, or one ^{3}H-atom per four bombykol molecules, about 10^9 molecules or 4×10^{-13} g were required for a measurement in the scintillation counter. The odour source, a 1cm^2 piece of filter paper (f.p.), had to be loaded with 3×10^{-12} g of bombykol in order to induce wing fluttering of some of the moth with a ten-s stimulus. Almost all of the responses occurred within two s. The threshold curve (in % of moths responding within the first two s) covered about 2 decades of stimulus

A. Gutiérrez and S. Marco (Eds.): Biologically Inspired Signal Processing, SCI 188, pp. 45–52.
springerlink.com   © Springer-Verlag Berlin Heidelberg 2009

load. Depending on temperature, the time of the day, and the animal origin the 50% threshold was reached at loads between 10^{-11} and 10^{-10} g/f.p. (Kaissling and Priesner 1970).

The fraction of molecules on the filter paper that was released per s, given an airflow of 100 ml/s, was determined with loads of 10^{-8} to 10^{-4} g/f.p. The fraction was 1/60,000 at 10^{-8} and 10^{-7} g/f.p. This value was extrapolated for the load at the behavioural threshold. In our setup the concentration of stimulus molecules decreased on the way from the outlet of the airflow system to the antenna. The fraction of molecules released from the filter paper that was adsorbed on the antenna was 1/150, determined with loads of 10^{-6} g/f.p., or higher.

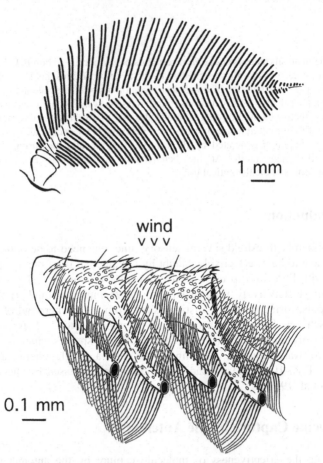

Fig. 3.1. Antenna of the male saturniid moth Antheraea polyphemus. Upper panel: Schematic view of the antenna. Each antennal stem segment has four side branches. Lower panel: Two antennal segments enlarged with different types of sensilla. The numerous, long olfactory hairs contain two or three receptor neurons responding to two or three components of the female pheromone.

With a load of 10^{-11} g/f.p. the concentration (c) of bombykol in air was 3,000 molecules/ml when the air stream velocity (v) was 57 cm/s as measured by means of a thermistor. Of the odour flow ($c \times v \times a$) (molecules/s) passing an area (a) equal to the outline area of the *Bombyx* antenna 27% was adsorbed on the antenna (Kaissling 1971). The air flow (air volume/s) through the actual antenna was most likely not more than 30% of the free air flow (through an area equal to the antennal outline area). A transmittance of about 30% of the free air flow may also be estimated for the much larger antenna of the male *Antheraea polyphemus* from measurements of air-stream velocity in front of and behind the antenna (Figures 3.1 and 3.2). The fraction

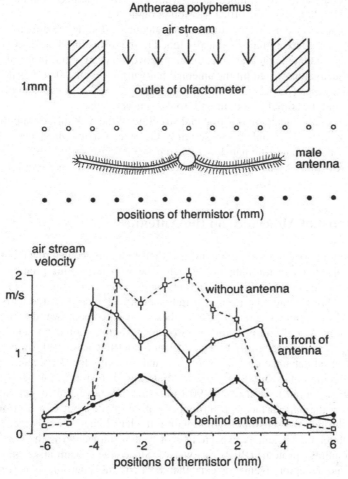

Fig. 3.2. Airflow at the antenna of the male moth Antheraea polyphemus. Air was blown from a glass tube towards the antenna. Scheme of the glass tube, antenna, and thermistor positions drawn to scale. By means of a thermistor (0.2 mm diameter) the airstream velocity was measured without the antenna (open sqares and dashed lines), in front of the antenna (open circles), and behind the antenna (dots).

of molecules passing an area equal to the antennal outline area that was adsorbed on these antennae was 32% (Kanaujia and Kaissling 1985). Thus we may conclude for both species of moths that from the air passing the antenna itself all pheromone molecules were caught.

In males of *Bombyx mori* (Steinbrecht and Kasang 1972) and *Antheraea polyphemus* (Kanaujia and Kaissling 1985) we determined the fraction of molecules caught by the antenna that was adsorbed on the long olfactory hairs (*sensilla trichodea*). The hollow, fluid-filled hairs are 2-3 micrometer thick and 100-300 micrometer long and house the sensitive dendrites of the pheromone receptor neurons. After 10-s exposure of single antennal branches to strong stimuli (10^{-4} g of ^{3}H-labeled pheromone/f.p.) the hairs were immediately (within 1 min) separated from the branch. 80% of the total radioactivity adsorbed was found on the hairs.

These findings show that the structure of the antenna, including the dimensions and arrangement of the olfactory hairs, is ideally tuned to the diffusion of odour molecules in air. It can be calculated that due to its thermal movements an odour molecule on its way through the antenna would hit the antennal hair surface about 100 times if it were reflected upon hitting. The design of antenna and hairs creates, as it were, an olfactory lens concentrating the stimulus and direct it to the sensory cells.

So far the exact chemical composition and structure of the hair surface and the pore tubules of the hair wall are unknown. Certainly the outer epicuticular layer is highly waterproof. If one damages the hair locally using a small laser beam one can - under microscopical control - see an air bubble growing starting from the point of hitting the hair.

3.3 Transport of Molecules on the Antenna

Following a strong concentration gradient, the pheromone molecules move along the hairs towards the body of the antennal branch. The velocity of this process can be measured if the hairs are cut at different times after exposure to ^{3}H-labeled pheromone. Within minutes the measured radioactivity decreased on the hairs while increasing on the antennal branch. From these measurements we determined a diffusion coefficient D of 50 μm^2/s for the movement of pheromone on the hairs of *B. mori* (Steinbrecht and Kasang 1972). The diffusion coefficient was 90 μm^2/s for air-filled hairs of dried antennae of *A. polyphemus*, and 30 μm^2/s for intact hairs of fresh antennae (Kanaujia and Kaissling 1985). Modeling diffusion in *A. polyphemus* (Kaissling 1987; unpubl.) we use D = 90 μm^2/s for the movement of the stimulus molecules along the hair surface and through the pore tubules, but D = 30 μm^2/s for the diffusion through the sensillum lymph within the hair lumen towards the receptor neuron. The latter coefficient is expected for a protein molecule of the size of the pheromone binding protein (PBP) in water. Since the quantitative model of perireceptor and receptor events reveals that 83% of the pheromone adsorbed is bound to the PBP within less than 3 ms, we can conclude that the longitudinal movement of the ^{3}H-labeled pheromone represents the movement of the pheromone-PBP complex (Kaissling 2001; unpubl.). The remaining 17% of pheromone molecules are enzymatically degraded inside the hair lumen and may no longer function as stimulants.

With the above-mentioned diffusion coefficients the modeled delay of the molecule arrival at the receptor cell is about 10 ms after adsorption at the olfactory hairs (Kaissling 2001, Figure 3.7B). This fits to the minimum delay of the receptor potential, the first bioelectrical response of the receptor neuron, as measured after stimuli of high intensity. At weak stimulation the average delay of the responses is a few hundred ms due to the chemical reactions of the stimulus molecules including their interaction with the receptor molecules (Kaissling 2001, and unpubl.).

3.4 Cellular Transduction

At low stimulus intensities about 25% of the pheromone molecules adsorbed on the antenna elicit nerve impulses of the receptor neuron (Kaissling 1987). Modeling reveals that – besides the 17% enzymatically rapidly degraded molecules - more than half of the molecules adsorbed must be lost due to the - still hypothetical - odorant deactivation on the hairs (Kaissling 2001). The 25% fraction of effective molecules was determined by radiometric measurements and by counting the nerve impulses at low stimulus intensities such that about one nerve impulse is elicited per receptor neuron by a one-s stimulus ($3 \times 10{-}10$ g of bombykol/f.p.). At and below this stimulus intensity one pheromone molecule is sufficient to elicit a nerve impulse (Kaissling and Priesner 1970).

The first responses of the receptor neuron to a single pheromone molecule are one or a group of small depolarizations (elementary receptor potentials, ERPs) (Kaissling and Thorson 1980; Kaissling 1994). The single ERPs with amplitudes of $0.1 - 1$ mV in extracellular recordings last about 10 ms and may trigger firing of one, seldom more than one nerve impulse. Quantitative modeling suggested that *in vivo* it is the odorant-PBP complex rather than the free pheromone which interacts with the receptor molecule (Kaissling, 2001). The ternary complex PBP-pheromone-receptor may one or several times turn into an active state before it finally dissociates (Minor and Kaissling 2003). Each activation causes – via an intracellular cascade of signal processes – a transient conductance increase of about 30 pS as reflected in the ERP.

3.5 Processing in the Central Nervous System

The extreme sensitivity of the receptor neurons is combined with a most efficient processing of their responses by the central nervous system. Via the axons of the receptor neurons the nerve impulses are conducted to the antennal lobe, the first synaptic station of the central olfactory pathway in insects. The axons of the pheromone receptor neurons terminate on local interneurons and projection neurons (PN) of the macroglomerular complex (MGC) (Hildebrand 1996). The silk moth has 17,000 bombykol receptor neurons per antenna (Steinbrecht 1970) and 34 projection neurons connecting the MGC with higher centres of the central nervous system (Kanzaki et al. 2003). Since there are also 17,000 bombykal receptors the messages of at least 1000 (in the hawk moth *Manduca sexta* up to 10,000) receptor neurons finally converge to one projection neuron.

The convergence of receptor neurons lowers the threshold of pheromone detection by integrating the nerve impulses (spikes). Since the receptor neurons occasionally fire nerve impulses without stimulation, this background activity produces the noise which needs to be distinguished from the signal, i.e. the stimulus-induced activity. The spikes of single receptor neurons were counted every 0.1 s for two s after stimulus onset. The background frequency (f_{bg}) in spikes/s counted after control stimuli with clean air was subtracted from the frequency counted at pheromone stimulation in order to obtain the stimulus-induced frequency (f_{st}). Most of the behavioural responses (wing vibration) started within the first two s after stimulus onset, with an average delay of 0.2 s, the integration time (t_i) of the central nervous system. Since the background frequency of the bombykol receptor neuron $(f_{bg} = 0.0855$ spikes/s) has a random distribution (Kaissling 1971), its variability represents the noise that determines the recognizability of the signal. The noise is proportional to the square root (sqrt) of f_{bg}. Also the stimulus-induced frequency (f_{st}) was shown to be randomly (Poisson) distributed at stimulus intensities eliciting less than three nerve impulses per stimulus (Kaissling and Priesner 1970). For a number (n) of receptor cells the noise is

$$\sqrt{n \cdot t_i \cdot f_{bg}} \qquad (3.1)$$

while the signal is

$$n \cdot t_i \cdot f_{st} \qquad (3.2)$$

and the signal-to-noise ratio is

$$f_{st} \cdot \sqrt{\frac{n \cdot t_i}{f_{bg}}} \qquad (3.3)$$

With a load of 10^{-11} g of bombykol /f.p. (10^{-10} g/f.p.) 40% (80%) of the males responded with wing vibration (at 21°C). At these loads we found $f_{st} = 0.0145$ (0.1545) spikes/s (from Tab. 2 in Kaissling and Priesner, 1970). For a convergence of 17,000 receptor neurons we find from Eq. 3.3 and with $f_{bg} = 0.0855$ spikes/s a signal-to-noise ratio of 3 (31). This shows that the processing in the CNS works near the theoretical limit.

The exact pattern of neuronal connections in the antennal lobe and the mechanism of signal/noise detection in the CNS are unknown. If we assume a minimum convergence, i.e. that each PN (directly or via interneurons) receives input from 1000 receptor neurons, the signal-to-noise ratio would be smaller than calculated above for an input from 17,000 neurons: For the 40% (80%) behavioural threshold we find a ratio of 0.7 (7.5). In this case the signal-to-noise ratio would be below the significant level of 3, at least for the 40% threshold. Consequently higher centres would need to contribute to the signal/noise detection, by converging the messages delivered from the PNs to higher-order neurons.

Finally it should be mentioned that the bombykol concentration in air at the 40% (80%) behavioural threshold was 3,000 (30,000) pheromone molecules/ml of air, at an airstream velocity of 57 cm/s. Interestingly these moths with 17,000 receptor neurons/antenna are almost as sensitive as a dog for (other) odorants (1000 molecules/ml).

Since a dog may have 10,000-fold higher numbers of receptor neurons than the moth, its thresholds could be 100-fold lower than the one of the moth. It could be even lower since the dog´s integration time is probably larger than the one of the moth. It seems clear that factors other than the number of receptor neurons are important for a high sensitivity, such as a high effectiveness of molecule capture and conveyance to the sensitive structures, or a low background activity of the receptor neurons. The low threshold in dogs suggests that as in the moth single molecules are able to produce nerve impulses.

Acknowledgment

For linguistic improvements I thank Dr. Ann Biederman-Thorson, Oxford.

Dedication

This paper is dedicated to Prof. Dr. Dietrich Schneider, who started the analysis of pheromone detection in the silkmoth, on the occasion of his 88th birthday.

References

Butenandt, A., Beckmann, R., Stamm, D., Hecker, E.: Über den Sexuallockstoff des Seiden-spinners Bombyx mori. Reindarstellung und Konstitution. Z. Naturforschung 14b, 283–284 (1959)

Hildebrand, J.G.: Olfactory control of behavior in moths: central processing of odor information and the functional significance of olfactory glomeruli. J. Comp. Physiol. A 178, 5–19 (1996)

Kaissling, K.-E.: Insect olfaction. In: Beidler, L.M. (ed.) Handbook of Sensory Physiology IV, vol. 1, pp. 351–431. Springer, Heidelberg (1971)

Kaissling, K.-E.: Chemo-Electrical Transduction in Insect Olfactory Receptors. Annual Review of Neurosciences 9, 21–45 (1986)

Kaissling, K.-E.: R. H. Wright Lectures on Insect Olfaction. In: Colbow, K. (ed.), Simon Fraser University, Burnaby, B.C., Canada, pp. 1–190 (1987)

Kaissling, K.-E.: Elementary receptor potentials of insect olfactory cells. In: Kurihara, K., Suzuki, N., Ogawa, H. (eds.) XI Int. Symp. Olfaction and Taste, pp. 812–815. Springer, Tokyo (1994)

Kaissling, K.-E.: Pheromone-controlled anemotaxis in moths. In: Lehrer, M. (ed.) Orientation and Communication in Arthropods, pp. 343–374. Birkhaeuser Verlag, Basel (1997)

Kaissling, K.-E.: Olfactory perireceptor and receptor events in moths: A kinetic model. Chem. Senses 26, 125–150 (2001)

Kaissling, K.-E., Priesner, E.: Die Riechschwelle des Seidenspinners. Naturwissenschaften 57, 23–28 (1970)

Kaissling, K.-E., Thorson, J.: Insect Olfactory Sensilla: Structural, Chemical and Electrical Aspects of the Functional Organisation. In: Satelle, D.B., Hall, L.M., Hildebrand, J.G. (eds.) Receptors for Neurotransmitters, Hormones and Pheromones in Insects, pp. 261–282. Elsevier/North-Holland Biomedical Press (1980)

52   K.-E. Kaissling

Kanaujia, S., Kaissling, K.-E.: Interactions of Pheromone with Moth Antennae: Adsorption, Desorption and Transport. J. Insect. Physiol. 31, 71–81 (1985)

Kanzaki, R., Soo, K., Seki, Y., Wada, S.: Projections to higher olfactory centres from subdividions of the antennal lobe macroglomerular complex of the male silkmoth. Chem. Senses 28, 113–130 (2003)

Kasang, G.: Tritium-Markierung des Sexuallockstoffes Bombykol. Zeitschr fuer Naturforschung 23b, 1331–1335 (1968)

Minor, A.V., Kaissling, K.-E.: Cell responses to single pheromone molecules reflect the activation kinetics of olfactory receptor molecules. J. Comp. Physiol. A 189, 221–230 (2003)

Schneider, D., Kasang, G., Kaissling, K.-E.: Bestimmung der Riechschwelle von Bombyx mori mit Tritium-markiertem Bombykol. Naturwissenschaften 55, 395 (1968)

Steinbrecht, R.A.: Zur Morphometrie der Antenne des Seidenspinners, Bombyx mori L.: Zahl und Verteilung der Riechsensillen (Insecta, Lepidoptera). Z Morph Tiere 66, 93–126 (1970)

Steinbrecht, R.A., Kasang, G.: Capture and conveyance of odour molecules in an insect olfactory receptor. In: Schneider, D. (ed.) Olfaction and Taste IV, Stuttgart, Wissensch Verlagsges, pp. 193–199 (1972)

4

Multivariate Analysis of the Activity of the Olfactory Bulb

I. Montoliu[1], K.C. Persaud[2], M. Shah[2], and S. Marco[1]

[1] Departament d'Electrònica. Universitat de Barcelona. Martí i Franquès, 1.
E 08028 Barcelona (Spain)
[2] SCEAS. The University of Manchester. PO Box 88,
Sackville St. Manchester M60 1QD, (United Kingdom)

Abstract. In the understanding of processes of neural activity in complex networks, non-invasive recording of the electrical activity is desirable. One method that achieves this is through the use of voltage-sensitive fluorescent dyes (VSD) as reporters, which convert changes in a tissue's membrane potential into fluorescent emission. This fluorescent activity is recorded by means of fast CCD cameras that allow visualization and a posteriori studies. Image sequences obtained in this way, although often noisy, are commonly studied following a univariate approach.

In this work, there are studied several series of image sequences from an experiment initially focused in the monitoring of the Olfactory Bulb activity of a frog (Rana *temporaria*), where the olfactory receptors were exposed to two volatile compounds, under different inhibitory conditions. Our work proposes the use of two multivariate analysis methods such Multi-way Principal Component Analysis and Independent Component Analysis with a dual aim. First, they are able to improve the recordings by removing noise and aliasing after using a supervised selection of parameters. Secondly, they demonstrate possibilities in the obtaintion of simultaneous information about the most active areas of the monitored surface and its temporal behaviour during the stimulus.

4.1 Introduction

In the Olfactory Bulb (OB), signals coming from the receptor neurons are initially processed, the output signals then going on to the olfactory cortex. Because of the nature of the signals and environment (usually 'in vivo') the study of functionality is not simple. With this aim, the monitoring of neural events in the olfactory bulb in real-time during the course of a olfactory stimulus has been chosen as a strategy to get more information about its behaviour. This monitoring is difficult, because often it is focused in the detection and temporal recording of changes in membrane potentials. The use of microelectrodes has its drawbacks, due to low magnitude of the signals, its invasive character and the difficulty of recording more that 2-3 cells at once.

In front of these problems, as an alternative to study the OB activity, it was proposed (Kent and Mozell 1992) the recording of images after impregnating the tissues with an appropriate dye.. These recordings should provide, not only a better spatial visualization of the global activity of the area under study but also the possibility of a non-invasive measurement. From this moment on, it would be possible to extract both temporal and

A. Gutiérrez and S. Marco (Eds.): Biologically Inspired Signal Processing, SCI 188, pp. 53–72.
springerlink.com   © Springer-Verlag Berlin Heidelberg 2009

spatial features elicited from a specific response and to record the overall response from several neurones at a time. To develop the technique it was necessary to combine neurobiology, spectroscopy, organic chemistry, optics, electronic, electro-optics and computing.

Voltage-sensitive fluorescent dyes (VSD) act as optical transducers of membrane potential changes. These compounds are capable of staining living cells, binding to the external cell membranes.. In this way, these compounds are able to respond to the membrane potential, without interfering in living cell's normal operation. This response is characterized by changes in the fluorescent emission and its wavelength. This fact must be considered previously to the imaging. Their response time towards potential changes is around a μs. Thus, these dyes provide the possibility of non-invasive multi-cell recording of the electrical activity. In this way, is possible to study the spatial-temporal structure of the activity in neuronal cell populations.

Finding the proper conditions of dyeing is not simple, because several undesired effects can occur. Intoxication of live cells due to the staining process can be possible if an excess of dye is applied. This excess can be the responsible of changes of membrane permeability, thus affecting the living cells. To apply the optimal concentration of the dye is necessary to consider the binding constant of the system dye/membrane. It is commonly accepted as a general rule to use the minimum possible amount of dye in each staining process. Intoxication is not the only side effect that can take place. Other effects, such the photoinduced effect must be taken into consideration. Basically, this effect consists in the release of oxygen singlets, highly reactive, from the dye molecule due to a light-induced reaction, caused by the intense light source. The presence of this reactive oxygen can damage the living cells, altering its functionality. One way to circumvent this effect is to do the measurement at different wavelengths to the ones presenting maxima in absorbance.

There are technological difficulties in the recording of images in these environments. It is necessary to have access to an ultra fast response device and suitable data acquisition devices, to keep the records in the same time scale of the observed activity. Recording pioneers of the activity of the OB (Orbach and Cohen 1983) after staining using VSDs, used an array of diodes as a detector with the aim of achieving very fast responses. Even though this fast acquisition was achieved, the dimensions of the array (12 x 12) or (16 x 16) were an important drawback. In this sense, the low resolution of the image became a loss in spatial information.

This work is based on recording of the activity of the OB of the frog (Rana temporaria), designed to investigate the inhibitory effect of lectin Concanavalin A (Con A) applied to the olfactory mucosa on the fatty acid response seen in the OB (Shah et al 1999). The olfactory mucosa was exposed to brief pulses of different odorants in the presence and absence of Con A, and the resulting responses follwed in the OB using a voltage sensitive dye as a reporter of neuronal activity. Two additional measurements were done for each analyte, corresponding to the observed activity after the removal of Con A with Ringer's solution. To ensure comparable results, the positioning of the camera has been thoroughly checked after 5x magnification in a small area containing 3-4 glomeruli, corresponding to the left side of frog's brain.

This chapter puts the emphasis in data processing techniques for the extraction of features and a simultaneous noise removal from images of brain activity. Two data analysis techniques are presented based in linear methods for dimensionality reduction. Its aim is double: on one hand, they can act as suitable techniques for image filtering/denoising. On the other, in the same step, they provide spatial-temporal information allowing the correlation between a specific area and its behaviour along the measurement time.

4.2 Methodology

4.2.1 Staining

As it has been described above, proper staining of the tissues is critical to obtain a reliable response. Among the population of voltage-sensitive fluorescent dyes, for this experiment it was selected RH 414 (4-(4'-p diethyl aminophenylbuta-1', 3'-dienyl) - γ-triethyl ammonium-propyl pyridinium bromide. This is a compound fluorescent-active when it is in the hydrophobic environment of cell membranes. It belongs to the family of styryl dyes, which have some advantages, such its good dynamic range, its low toxicity and a low level of photoinduced damage potential.

A solution of $1mg\ ml^{-1}$ of this dye in Ringer solution was prepared for the staining of the tissues. The olfactory bulb of the frog was exposed. To properly stain the olfactory bulb, 7μl of solution were applied drop-wise onto the surface to be stained, covering the whole area. During the staining period (between 30' and 45'), the cavity was covered with a small glass.to avoiding loss of moisture during incubation. Any excess of staining solution was then removed using Ringer's solution as solvent, and excess solution removed using the capillary action of a small piece of tissue paper. After this step, the tissue was considered to be ready for imaging.

Once the tissues were ready, it was necessary to build an optical imaging system that can be described at 3 different levels: optical, photo detection and data acquisition.

4.2.2 Optical Section

In the recording of the images, an epifluorescent metallurgical microscope (Olympus, Japan) was used. This instrument was appropriately modified with a set of lenses and filters, and placed after a light source of 12V/100W. In particular, three filters of different types were installed: an interference excitation filter, a barrier filter and a dichroic filter. Both three allowed the separation of excitation/emission wavelengths during the imaging. Thus, it was possible to discriminate between the excitation radiation of λ_{exc} 520-530 nm and the fluorescent emission of λ_{em} 600 nm. The whole apparatus was placed in an optical vibration-isolated table, within a Faraday cage, to obtain an electromagnetically noise free environment. The optical instrument was preset to operate allowing 20 x magnifications.

Photo Detection

As mentioned above, to get the images at the necessary speed during the experiment, it is mandatory to have a fast camera. In this case, fast image detection was confided to a CA-D1-0064 Turbosensor Area Scan Camera (DALSA Inc. Canada). This system consists in a high-speed monochrome camera with a CCD array of 64x64 pixels

(1mm^2 each). The CCD sensor has adjustable pixel frequencies of 4, 8 or 16 MHz. Together with the data acquisition system (see below), they provide recording times of 1160 ms, 580 ms and 290 ms, respectively. In the image series under study, the pixel frequency was set to 4 MHz, thus leading to a 1160 ms of full sequence capturing in 832 frames. This leaded to an integration of 1.39 ms of response for each frame. After 20x optical magnification, the effective recording area is 63x63 pixels.

Data Acquisition

To convert the information recorded by the CCD device and to be stored in a suitable data format, a 4 MEG Video Frame Grabber (model 10, EPIX, USA) was used, together with a DALSA camera interface card (EPIX, USA). The grabber was synchronized with the camera, to achieve a 'in phase' mode of operation, and was able to operate at pixel speeds up to 19MHz. The DALSA camera interface card was able to digitise with an 8-bit resolution, thus rendering up to 256 grey levels. This device is able to digitalize video frames of 80 x 63 pixels, and up to 832 frames. The maximum amount of frames is derived from the device's buffer size (4096 Kb) and the image size (5040 bytes). This system was connected to an IBM-PC compatible system, to act as a data logger and to a RGB monitor, also controlled by the frame grabber, to enable carefully positioning of the camera over the OB region.

Properly synchronization between odour stimulation and imaging was enabled. Data acquisition was delayed 150ms after odour delivery. It was done so, as a trade-off between the time of response to an odorant stimulus (described to be around 100ms) and also the number of frames available. The odorant pulses were delivered to the left nasal cavity of the frog, through a valve-operated system, in 60 ms air pulses.

In this way, six movies were generated. They corresponded to stimulation using Butyric acid and Isoamylacetate before, after the addition of Con A and after further removal of the lectin. Its distribution is shown in table 4.1.

Table 4.1. Labelling of image series

Odorant	Direct	After Con A	After rinsing with Ringer Solution
Butyric Acid	R040212F.97	R060212F.C97	R110212F.R97
Isoamyl Acetate	R020212P.97	R080212P.C97	R120212P.R97

4.2.3 Data Processing

Some considerations about the nature of information in a movie, from the data processing point of view are important. Conceptually, a movie is a set of frames recorded sequentially. Formally, it is a dataset that presents some structure due to the sequential nature of its process of generation. Thus it means that the different frames are ordered in time, and each frame consists of an image with X and Y-axis. These datasets, in which there exists a special ordering due to external reasons (often time in Statistic Process Control), are considered multi-way datasets. In our case, we can assume that we are dealing with a three-way dataset, because it presents a structure in the way (time x X_pixel x Y_pixel). In this way, (I, J, K) axes are providing different information. The most obvious slice (J x K), corresponds to the square image recorded and it is ordered sequentially along the I direction (time). The other two possibilities (I x J, I x K), describe the variation of all the pixels of one axis along the other axis, for all the time period. Whilst the first case has a physical meaning, because it describes the spatial conformation of the images, in both of the latter cases this information does not have a practical interpretation.

Multi-way Principal Component Analysis

Multi-way Principal Component Analysis (MPCA) is strongly related to the standard data analysis method Principal Component Analysis (PCA). This bilinear modelling technique, based in the eigenvector decomposition of the covariance matrix, does not consider the way in which data has been acquired. This means that external information, such the ordering in time of the data acquisition, is not taken into account for the modelling process. Although this is unnecessary in a wide amount of cases, there are some for which it becomes an evident loss of information. Multi-way are part of these.

Multi-way PCA is statistically and algorithmically consistent with PCA (Wise et al. 1999; Westerhuis et al. 1999). Thus, it decomposes the initial matrix $\underline{\mathbf{X}}$ in the summation of the product of scores vectors ($\mathbf{t}$) and loading matrices ($\mathbf{P}$), plus a residual matrix ($\underline{\mathbf{E}}$). These residuals are minimized by least squares, and are considered to be associated to the non-deterministic part of the information. The systematic component of the information, expressed by the product ($\mathbf{t}$ x $\mathbf{P}$), represents

$$\mathbf{X} = \sum_{i=1}^{k} \mathbf{t}_k \otimes \mathbf{P}_k + \mathbf{E} \tag{4.1}$$

the deterministic part of the data. This part is decomposed, in our specific case, in a scores vector related to the variation of the data response detected at each sampling time and a loadings matrix, corresponding to the spatial distribution of the activity. This decomposition, as in PCA, is ordered in decreasing order of importance, considering the directions of maximum variance. In this way, MPCA is equivalent to carrying out PCA onto an unfolded data array across one of the modes, for instance (I x JK) (Figure 4.1). In practice, this means a vectorization of each frame and its' ordering along recording time. Even though there exists other kind of unfoldings, this one allows summarizing the variance contained in each frame at each recording, and isolates the time-related information from the spatial component.

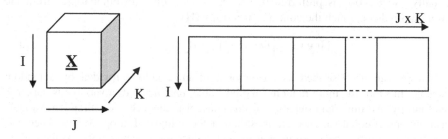

Fig. 4.1. Data handling of Multiway arrays. Unfolding along one dimension.

In brief, the MPCA algorithm correctly unfolds the data matrix $\underline{X}$, keeping the I dimension constant. This, now bivariate, matrix X is mean-centered for each variable value (J x K variables) and auto scaled also to unit variance if necessary. Eigenvector decomposition is applied to the covariance matrix of X, with the aim of obtaining t and p. In this decomposition, it must be remarked that p is, in fact, an unfolded version of P_k. This last matrix is easily obtained by proper folding of p matrix, while residuals matrix $\underline{E}$ is obtained also in a similar way, after proper folding of an equivalent bivariate e matrix. The nature of the eigenvector decomposition of the covariance matrix, points out one of its differential characteristics. Thus, MPCA decomposes the information in directions determined using a criteria of variance maximization, keeping a restriction of orthogonality between directions in the, now reduced, multivariate space.

Independent Component Analysis

Linear transformation of the original variables can lead to suitable representations of original multivariate data. As is shown above, MPCA method makes this transform pointing towards directions of maximum variance. In Independent Component Analysis (ICA) the goal is finding components (or directions) as independent as possible. This linear decomposition of one random vector (multivariate data) x follows the expression:

$$x = As \qquad (4.2)$$

where A is the so-called mixing matrix and s the independent component vector. This is a generative model, meaning that it describes how data vector x has been generated by a mixing process (A) of the components s. Because both A and s are unknowns, they must be generated from x after applying some assumptions. In ICA, the statistical independence between components s_i must be ensured.

Independence of two variables is related to the fact that one variable does not give information related with the other and vice-versa. This concept can be described in terms of probability density functions. In this case, independence between two variables can be assumed if the joint probability density of these two variables factorizes. This condition is achieved if the data are non-Gaussian. Consequently, non-Gaussianity of the data can be considered a good indicator of independence (this is so because, in Gaussian variable systems, the directions of the mixing matrix A cannot be determined if the independent components are also Gaussian). Non-gaussianity can be measured in two ways: using Kurtosis and Negentropy (J) criteria. Usually, Negentropy is preferred due to its connection with other topics from the information theory, such the concept of Entropy (H).

$$H(y) = - \int f(y) \log f(y) dy \qquad (4.3)$$

Entropy can be described as the amount of information provided by a random variable. In short, the more random (unpredictable, unstructured) a variable is, larger is the Entropy. An important outcome of this concept is that (Gaussian) random variable distributions of equal variance are those that achieve higher Entropy values. Then it is possible to identify a decrease in entropy with a decrease in gaussianity. At the end, to obtain a positive value for non-gaussianity, is defined the concept of Negentropy (J).

Negentropy consists in the difference between the maximum Entropy for this variable (defined as Gaussian random variable of the same covariance matrix) and its current Entropy. Thus, this magnitude always achieves a positive value and becomes an optimal estimator of non-gaussianity. Unfortunately, the estimation of Negentropy is difficult and some approximations must be done to calculate its value.

In this sense, there are some simplifications in the literature. They lead at the end to a very simplified expression of Negentropy that can depend from only one non-quadratic function G, given by:

$$J(\mathbf{y}) \propto \left[E\{G(\mathbf{y})\} - E\{G(v)\} \right]^2 \tag{4.4}$$

Where $\mathbf{y}$ is a mean-centered and variance scaled variable. Appropriate selection of G can provide better approximations of Negentropy. Some examples of selected G are:

$$G_1(u) = \frac{1}{a_1} \log \cosh(a_1 u), G_2(u) = -\exp\left(-u^2/2\right) \tag{4.5}$$

Where $1 \leq a_1 \leq 2$ is a constant to be selected.

Another way of finding the suitable estimations of independence is obtained through the minimization of mutual information. In brief, mutual information can be defined as a measure of the dependence between random variables. It is defined by:

$$I(y_1, y_2, \dots y_m) = \sum_{i=1}^{m} H(y_i) - H(\mathbf{y}) \tag{4.6}$$

Its value is always nonnegative and becomes 0 if the variables are statistically independent. In opposite to PCA and related methods, which makes use of the covariance to reflect data structure, mutual information takes into account the whole dependence structure of the variables.

As in other methods, some pre-processing steps can be applied before analyzing data with ICA. In this sense, two pre-processing steps are assumed as standards. Basically, mean centering as in PCA and related techniques and whitening of observed variables. This last pre-treatment is equivalent to perform an eigenvalue decomposition of the covariance matrix. In this way, is possible to obtain a new matrix of whitened vectors, now uncorrelated, with unitary variance. The advantage of using this procedure is that an orthogonal mixing matrix $\mathbf{A}$ can be obtained. This fact reduces the number of parameters to be estimated more or less by a half, and leads to a simplification of the ICA algorithm. In this whitening step there is also performed a dimensionality reduction. This is done taking in consideration obtained eigenvalues and discarding those that are too small.

Once introduced objective contrast functions for ICA estimation, is necessary to use a maximization algorithm for the function selected. This maximization can be achieved using different procedures. In this work, we have selected the FastICA (Hyvärinen et al. 2001) algorithm. Among several properties of this algorithm there can be remarked: its quadratic convergence, thus guaranteeing fast resolution of the models; it is not a gradient based algorithm, thus avoiding the need of fixing step size as in linear perceptrons; its performance can be tuned checking different G functions and the independent components can be calculated one by one, thus avoiding finding unnecessary components in exploratory analysis. Even though there are proposed some specific functions in the literature, the algorithm can deal with almost any non-Gaussian distribution using any nonlinear one.

Describing FastICA for one single vector $\mathbf{x}$ and one single independent component, this algorithm finds a direction, described by a weight unit vector ($\mathbf{w}$) such the projection $\mathbf{w}^T\mathbf{x}$ maximizes non-gaussianity. Then, non-gaussianity is evaluated using the approximation of Negentropy $J(\mathbf{w}^T\mathbf{x})$ given above, on whitened data. This maximum is found following Newton's approximate method. It must be considered g as the derivative of the non-quadratic functions G mentioned above. Basically, FastICA does the following steps:

1. Random generation of one vector $\mathbf{w}$.
2. Estimation of an update of the vector $\mathbf{w}$:

$$\mathbf{w}^+ = E\{\mathbf{x}\,g(\mathbf{w}^T\mathbf{x})\} - E\{g'\left(\mathbf{w}^T\mathbf{x}\right)\}\mathbf{w} \qquad (4.7)$$

3. Normalization of $\mathbf{w}$:

$$\mathbf{w} = \mathbf{w}^+\big/\big\|\mathbf{w}^+\big\| \qquad (4.8)$$

4. Loop until convergence of $\mathbf{w}$.

The generalization of this algorithm for estimating several independent components operates in a similar way, but now considering an array of weight vectors $(\mathbf{w}_1,...,\mathbf{w}_n)$. In this case, it is necessary to prevent the convergence to the same maxima for different vectors. This can be achieved after a de-correlation step on the different outputs $\mathbf{w}_1^T\mathbf{x},...,\mathbf{w}_n^T\mathbf{x}$.

In brief, ICA provides a linear model of the information, decomposing it in different contributions under criteria of independence maximization. These directions in the space are described by the independent components, which can be considered formally equivalent to PCA loadings. The mixing matrix provides the contribution of each random vector to each direction in the independent components space, which is formally equivalent to PCA scores.

4.3 Results

After importing the data from the original format by using a suitable routine, the data were ordered accordingly to obtain its visualization. Before performing any analysis, it was necessary to overcome the differences due to the area covered by the frame grabber (ready to accommodate a rectangle of 80 x 63 pixel size) and the area covered by the camera (63 x 63 pixels). For each of the imaging series, these differences were solved simply removing the non-active area of the image. After that, it was possible to obtain a 3-way data array of 832 x 63 x 63 dimensions.

Analysis of data was performed in Matlab 7.1 sp3 (The Mathworks, USA) on an AMD64 Dual Core platform running on Linux. Standard routines contained in the PLS Toolbox 3.5 (Eigenvector Research, USA) and FastICA 2.5 (Hugo Gävert, Jarmo Hurri, Jaakko Särelä, and Aapo Hyvärinen, Finland) were used to build the linear models. Data handling, manipulation and representation were done using appropriate scripts written also in Matlab.

4.3.1 MPCA Analysis

Data were arranged in order to identify the time axis along the sample mode of the algorithm (in MPCA, by convention, the k-axis). Before calculating the principal components, mean centering of the variables obtained after the unfolding of each frame was performed. After this pre-processing step, a principal components model is set up from these data. Due to the noisy view of the images and other sampling artefacts such aliasing, several tests were done to decide a minimum of variance to be explained. Thus, a variance threshold was fixed at around 70% for each MPCA model. These variances were described by 15 principal components, and provided the most significant information in terms of contribution to the total variance.

As is known, PCA related methods provide an orthogonal decomposition of the covariance matrix after the solution of an eigenvalue problem. These results are usually sorted according to eigenvalues as indicators of the importance of each eigenvector on the global solution. This eigenvalue / eigenvector pair provides an ordered sort of the so-called principal components, in decreasing order of importance. This means, that each of the presented factors are contributing with different extension to the global data set.

The results obtained are shown in table 4.2. As it can be seen, between a 70 and 82 % of the total variance is being described by the 15 components of the MPCA model. A special case is given by the image series R060212F.C97, corresponding to the addition of the Con A to the OB to inhibit the response against Butyric acid. In this case, only 61.5% of the total variance can be explained with this number of components.

Table 4.2. MPCA variance results. Selection of components to be kept in the reconstruction with filtering purposes. Amount of variance selected from the total.

Sample Name	Factors selected	Total Variance (%)	Variance selected (%)
R040212F97	1, 4, 10	70.30	28.02
R060212F.C97	none	61.54	-
R110212F.R97	1, 4, 7, 8	75.66	40.06
R020212P.97	1, 6, 7, 8, 14	79.61	44.50
R080212P.C97	1, 4, 7	84.26	60.09
R120212P.R97	1, 4, 7, 12	75.57	43.94

In general, it can also be said that the magnitudes of the modelled variances are relatively low, which can be related to high noise content. It can also be observed a greater amount of variance for the three last image series, corresponding to

Isoamylacetate stimulation. This fact can be related to a greater intensity directly linked to the activity shown in the images. The origin of this intensity can be diverse, because it can be interpreted as a bigger response of the tissues towards Isoamylacetate stimulation, or it can be merely related to a more successful process of staining of the OB. Assuming that the staining process has been done at the same extent, this seems to reveal a more intense activity in the response to Isoamylacetate.

Factors Selection

Among the overall variance modelled in the first 15 components, a factor selection must be done. As expected, the goal of this selection is to obtain noise and aliasing free sets of images for their further interpretation. Therefore, the selected number of factors will define, as much as possible, the spatial-temporal patterns present in the data. To do it so, the scores (T) and refolded loadings (P) of each factor are studied. In this way, scores' interpretation provides a description of the temporal behaviour of the different activity patterns present in the data. These image activity patterns are described by the loadings of the model. To provide higher interpretability, a proper refolding of P to convert the vectorized image loadings (3969 elements) in a square-loading image of (63 x 63) pixels was done. As it can be seen on Figure 4.2, corresponding to the analysis of sample R040212F.97, the image loadings of the 15 components provide different patterns of activity. Some of them, corresponding to factors 1, 4, and 10 provide clear, noise and aliasing free, loading sets. These sets of selected scores and loading pairs (let's say T' and P') now can be used to build an enhanced version of the initial file. Therefore, the movie can be reconstructed

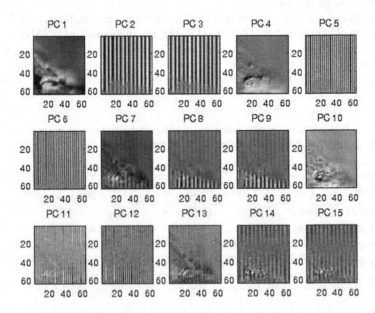

Fig. 4.2. Butyric acid MPCA loading images. Spatial patterns of activity identified at PC 1, 4 and 10. Identification of noise and aliasing effects.

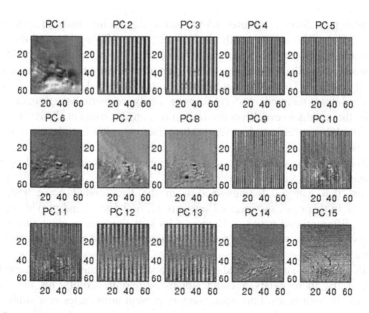

Fig. 4.3. Isoamylacetate MPCA loading images. Spatial patterns of activity identified at PC 1, 6, 7 and 8. Identification of noise and aliasing effects.

appropriately by doing the outer product of **T'** and **P'** matrices. This process provides a filtered imaging of the recording, as it can be seen in Figure 4.4. The reconstructed signal after MPCA analysis and posterior factor selection allows the identification of spatial activity patterns that were not available from raw signals. Although this method provides images which are almost clear, there are still present some minor aliasing interferences.

Having a look at the temporal patterns corresponding to the selected components is possible to see how there exist a counter-phase oscillatory behaviour for components

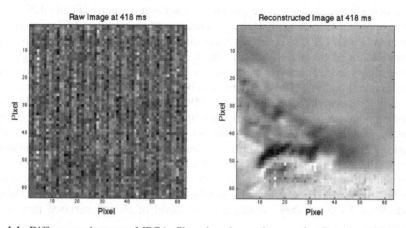

Fig. 4.4. Differences between MPCA filtered and raw images for Butyric acid response (R040212F.97). Selected frame at 418 ms.

4 and 10 before 594ms (Figure 4.5A). After that point, both factors reveal a synchronization of the activity with a sudden increase. Having in mind the variance ordered characteristics of the components obtained, it can be said that MPCA reveals three sources of oscillatory activity with differentiated amplitudes and contributions to the total variance. To check its relation with the functional activity of the OB, the two most relevant temporal patterns can be easily converted to frequencies. The analysis of these values reveals oscillations in response consistent (5 -10 Hz) with the stimulation-inhibitory effect observed at the olfactory epithelium level, as described in the bibliography.

After following the same process of refolding of the results obtained by MPCA, the analysis of Isoamylacetate images (R020212P.97) provide up to four principal components that can be distinguished from the background noise and aliasing (Figure 4.3). In this case, the selected ones are (1, 6, 7 and 8) and provide up to four regions of differentiated activity. Having a look at them, it is clear that some are providing very similar patterns. In particular, factors 6, 7 and 8 provide a high degree of similitude.

As in the previous case, temporal patterns are studied. In this way, as it can be seen in Figure 4.5B, there exists a counter-phase temporal activity for the first two components. In both cases, both score vectors present amplitudes in a similar range. The remaining factors (7 and 8) are presenting also temporal profiles of similar amplitude between them, but both are considerably lower in comparison with the preceding ones and are also noisier. Their contribution to the overall signal can be considered less important, and also agrees with the high similitude detected after loadings' analysis.

The analysis of the information contained in the imaging for butyric acid (after addition of Con A) and before the rinsing of this lectin with Ringer solution, leads to results according to what is expected. Thus, the addition of Con A, suppress almost completely the activity of the OB. The MPCA model for this imaging shows undefined spatial patterns showing noise and aliasing effects and reveal a unique activity zone in the lower part of the image corresponding to a blood vessel. Temporal patterns analyses agree with previous information and define constant profiles, without periodic oscillations, corresponding to the observed noise (Figure 4.5C).

After applying Con A to the OB, the analysis of the imaging of Isoamylacetate stimulation (R080212P.C97) presents some differences (see Figure 4.6). In this case, the suppression of the activity due to the Con A is not effective. Imaging of the response shows evident activity and posterior MPCA analysis provides differentiated spatial-temporal patterns. These patterns can be distinguished from the ones recorded previously for Isoamylacetate. In this case, MPCA analysis provides at least 3 loading images/temporal profiles describing the activity of the OB after the stimulus in these inhibitory conditions. They correspond to principal components 1, 4, and 7. Comparing those areas with the previous ones, corresponding to the non-inhibited response (Figure 4.3), they provide a certain level of simililitude although there are some minor differences, that can be related to the repositioning of the camera after depositing the lectin. There exists also a reduction of the number of factors to take in consideration. Even so, they contain a higher level of overall variance than in the previous case.

Butyric Acid   Isoamylacetate

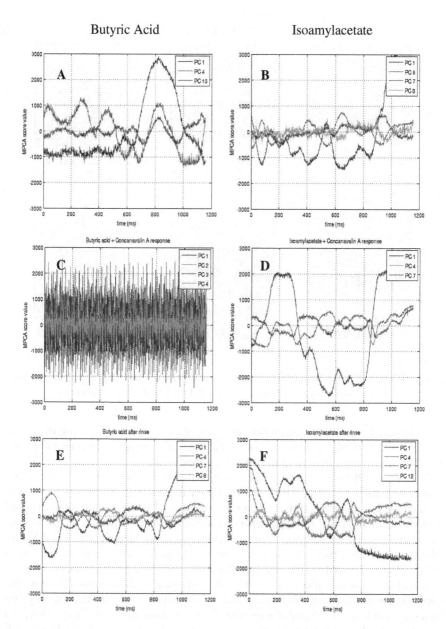

Fig. 4.5. MPCA analysis. Temporal patterns of Butyric acid and Isoamylacetate. Score vectors for direct stimulus (A,B), during inhibition with lectin (C,D), and after removal of the inhibitor (E,F).

There can be observed changes in the temporal profiles. As seen in Figure 4.5D, two of the components shown (4, 7) are also showing an oscillatory behaviour in counter-phase. Superimposed on both, there is presented a first component with

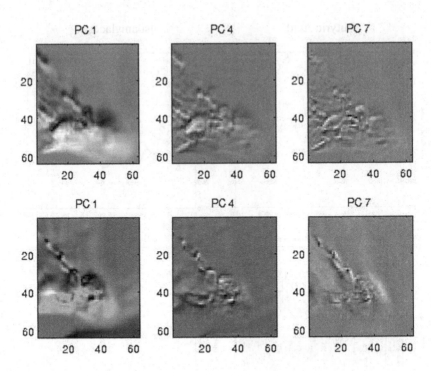

Fig. 4.6. Selected PCs for Isoamylacetate plus Con A (R080212P.C97). (Top row). Selected PCs for Butyric acid after rinse with Ringer solution (R110212F.R97). (Bottom row).

higher amplitude and lower frequency, and describes an oscillation along the entire recording. From these profiles, it can be said that the images are presenting two different sources of oscillation in the inhibitory conditions.

The rinsing of the Con A with Ringer solution should lead to a recovery of activity of the OB in those cases in which the inhibition is effective. According to what is expected, MPCA analysis on these data reveals renewed activity after being exposed to Butyric acid (R110212F.R97). Results obtained show in Figure 4.5E up to four clear spatial patterns of activity, present in principal components (1, 4, 7 and 8). As in the previous case, there appear three factors (1, 4, 7) with clearly differentiated spatial patterns of activity. Thus, the first three selected principal components (1, 4, 7) are providing a counter-phase oscillation. As expected, both three components are contributing with different amplitudes, due to the amount of information they are explaining. Although the third component is minor in comparison to the first two ones, it is clearly describing another source of oscillatory behaviour. The fourth component selected, namely factor (8), also shows a similar behaviour. However, it cannot be considered clearly relevant, due to its intensity (only comparable to principal component 7) and to the soft aliasing pattern (still present).

In comparison with spatial patterns for first response to butyric acid (R040212F.97) there appear some differences in general. An upward displacement of all the spatial activity is observed, that can be related somehow to the variability associated with the positioning of the camera after rinsing the tissues. Their temporal profiles still show a clearly definite oscillatory behaviour. But, in this case, the activity profiles appear to be slightly less intense as in the first stimulation, what can be interpreted as a general decrease in the activity after the Con A deposition / rinsing process. By the other side, the temporal activity profiles obtained by MPCA seem to be more definite and less noisy than the first stimulation, which points out to better recording conditions.

After rinsing the Con A and further stimulation with Isoamylacetate, it is possible to see changes in the temporal activity profile. As it can be seen in Figure 4.5F, these changes do not excessively affect the intensity of the temporal profiles, which remain almost in the same range. This similitude in intensity can be related to the ineffectiveness of the inhibition with the lectin that leaves mostly unaffected the response of the OB in terms of intensity. However, it is clear that there exist differences in the temporal response to the stimulus.

4.3.2 ICA Analysis

To matricize each of the three way datasets for ICA analysis, each of the movies was correctly unfolded doing an image vectorization on the pixel dimension. This operation rendered a 832 x 3969 data matrix for each of the experiments. Up to four nonlinear functions $g(u)=u^3$, $g(u)=tanh(u)$, $g(u)=u*exp(-*u^2/2)$ and $g(u)=u^2$ were tested for the calculation of the negentropy. After tests on each of the movies, for each of the possible nonlinear functions, it was observed that all of them were providing nearly the same models in the same conditions. Stability for some of the solutions and non-convergence in some cases, were used as criteria to select *tanh* as nonlinear function for all the models. Before doing any calculation, data were properly reduced in dimension and whitened to 15 principal components by means of PCA. After this step up to 15 independent components (IC) were calculated in parallel. Once the ICs were obtained, they were properly refolded to obtain the corresponding IC image and to identify spatial activity patterns. Mixing matrix (**A**) results were used to show the temporal patterns corresponding to the ICs. Contrary to PCA, in which there is possible an ordering of the PCs according to explained variance; in ICA this ordering is done with visualization purposes only. There are given some criteria in the literature to order the ICs in some way, such ordering according to eigenvalues obtained during dimensionality reduction and the use of reconstruction error as an indicator of the explained variance at those IC (Xueguang et al. 2006).

IC Selection

As it could be expected, the analysis of the ICs for both Butyric acid and Isoamylacetate stimulation reveals differentiated spatial-temporal patterns during the imaging. As it can be seen in Figure 4.7 for Butyric acid stimulation, ICA is able to isolate the contribution of 3 IC images (labelled IC 8, 13 and 14) from the global signal. The method is, then, capable of separating these contributions from aliasing and noise effects, clearly presents in the remaining ICs. The different spatial patterns of activity are concentrated around the same spatial region, corresponding to the glomeruli in focus.

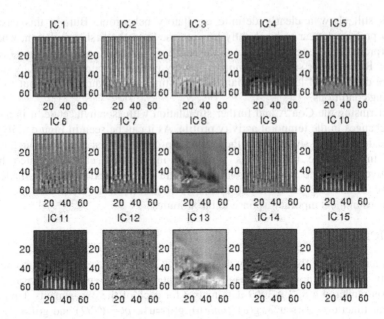

Fig. 4.7. Butyric acid IC images. Spatial patterns of activity identified at IC 8, 13 and 14. Identification of noise and aliasing effects.

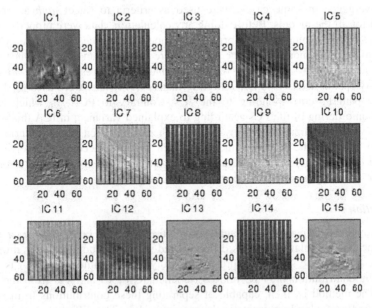

Fig. 4.8. Isoamylacetate IC images. Spatial patterns of activity. IC selected: 1, 6, 13 and 15.

For Isoamylacetate stimulation is also possible to isolate 4 IC images (marked as IC 1, 6, 13 and 15 in this case), leaving a remaining sort of IC images related to noise plus aliasing effects (Figure 4.8). The origin of these differences between imaging due to analyte stimulation is not clear from these ICs. Even though they could come by differences in OB response towards different analytes, it must be noted that these images correspond to two individuals, which can provide some morphological diversity, and consequently less reproducibility on the spatial patterns.

Inspecting the mixing matrix results, descriptors of the temporal activity pattern of the different IC images, it is possible to identify differentiated behaviours for the three regions. When these profiles are superimposed (Figure 4.9B), it can be seen a counter-phase oscillatory behaviour along all the 1160ms. The analysis of the mixing matrix vectors corresponding to the remaining non-selected ICs reveals both the noise contributions to the independent components and also the aliasing effects, present along the recording time with similar intensity. They also reveal how some of the information is still contained in those ICs, showing the limitations of the proposed method.

When the results obtained for both analytes are compared, it is possible to detect differences between both models. The mixing matrix results for Isoamylacetate provide differentiated temporal profiles for each of the selected ICs for the reconstruction. In this case, in Figures 4.9A and 4.9B, there can be observed a similar behaviour in frequency in the first 695 ms for both analyte stimulations. The biggest difference between both profiles appears after 695 ms, with the disappearance of the broad band present in Butyric acid temporal pattern. The inspection of the temporal profiles provides also criteria to discard some of the IC images for isoamylacetate, according to its relevance. In this way is possible to discard IC 6 due to its low oscillatory behaviour until 880 ms.

The analysis of the information contained in the imaging for butyric acid (after addition of Con A and before rinsing of this lectin with Ringer solution) leads to results according to what was observed in MPCA. The addition of Con A suppresses almost completely the activity for the OB and leads to an ICA model very similar in spatial-temporal activity to the one observed in the MPCA model. Thus, the ICA model also identifies noise and aliasing as the main contributors to the overall activity.

After the analysis of the imaging of Isoamylacetate stimulation under inhibitory conditions (R080212P.C97), ICA also reveals the differences observed in MPCA. The suppression of the activity due to the Con A is not effective. ICA analysis also provides differentiated spatial-temporal patterns that can be distinguished from the previous ones. Thus, the temporal profiles show changes in frequency and amplitude during the deposition and after removal of the lectin. As in MPCA, ICA is able to detect these changes, but considering the different sources of variation as purely independent. In this case, it provides up to 6 IC images/temporal profiles, almost free from noise and aliasing. The selection of the independent components is done in this case by dual inspection of the IC images and their mixing values. This inspection allows a selection to be made based in the dissimilitude of the observed image / temporal pairs. In this way, the former set of 6 IC images is reduced to 4 (4, 5, 8, and 12), as shown in Figure 4.10. The almost constant temporal activity of IC 12 and the identification of its associated spatial pattern reveal a strong differentiation for this area.

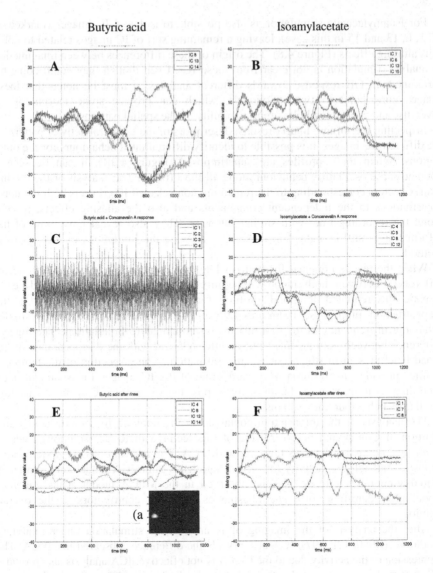

Fig. 4.9. ICA analysis. Temporal patterns of Butyric acid and Isoamylacetate. Mixing matrix vectors for direct stimulus (A,B), during inhibition with lectin (C,D) and after removal of the inhibitor (E,F). IC 8 image corresponding to damaged area (a).

After rinsing of the Con A, the recuperation of the initial behaviour is expected. ICA analysis on these data shows a recovery of the activity of the OB when it is exposed to Butyric acid (R110212F.R97). Although some activity is observed, results obtained show some differences in the IC images, corresponding to changes in spatial activity. As observed in MPCA analysis, an upward displacement of the entire activity pattern is observed. It is also observed in Figure 4.10 some strongly differentiated activity, described by IC 8. In this case, both the nearly constant temporal activity, seen in Figure 4.9E, and the spatial pattern of the surroundings point out to a damage of the tissue during the rinsing process. Three more ICs images are observed (4, 12, 14), which are related to the normal activity of the OB under Butyric acid stimulus. Whilst one of them keeps a certain degree of similarity with first IC analysis of the pure response, the remaining ones are showing a more uniform spatial pattern. When temporal patterns are studied and compared, before and after inhibition (and posterior rinsing), there can be observed relative changes in terms of relative intensity and frequency in each of the selected ICs as in MPCA. Changes in response are specially marked after rinsing. These observations points out to a change of the spatial-temporal activity of the OB after all the inhibition + rinsing process. The origin of these changes is diverse, and can be due to the progressive damaging of the tissues during the experiment (photoinduced damage and toxicity of the staining compounds), and also to small changes in the positioning of the camera.

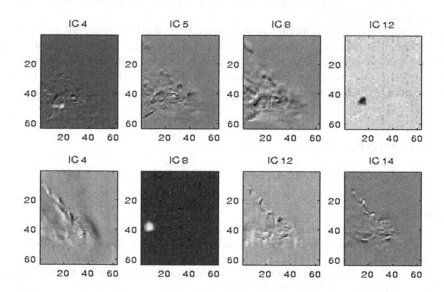

Fig. 4.10. Selected ICs for Isoamylacetate plus Con A (R080212P.C97). (Upper row). Selected Ics for Butyric acid after rinse with Ringer solution (R110212F.R97). (Bottom row).

4.4 Conclusions

In this work there have been presented two alternatives for data processing of image series, based in Multi-way PCA and Independent Component Analysis. Results obtained show that they can be used with a dual purpose. On one side, they can be used with the aim of filtering, removing noise and sampling artefacts, such aliasing. Another purpose, also important, is that they allowed the identification of the most active areas along all the recordings and its behaviour along all the recording time. This can permit the extraction of functional information.

The MPCA method is able to build a model of the spatial-temporal activity detected in the imaging. Because of the nature of the model, is possible to index the contribution of different components to the global activity. Consequently, it is possible to determine the weight of the patterns of activity along the entire recording. On the contrary this ordering cannot be easily achieved by ICA.

ICA is also able to build a model of the activity of each of the imaging series. The method is able to detect unexpected effects that probably follow an independent behaviour. In this way, is possible to detect some damage on the surface of the OB after the application / removal of Con A. These effects cannot be determined by MPCA.

Using these methods, it has been possible to determine changes in the overall response of the OB along the experiments. These changes appear to be due to the effect of the lectin, to the aging of the sample during the preparation and to the damages induced by the dye (photoinduced, toxicity). Temporal profiles obtained before and after the application of the Con A reveals that the recovery of OB's activity is not complete.

As expected, precise centering of the image onto the OB glomeruli is critical. Improvement in camera positioning and stimulation on the same individual with different analytes, should make possible to identify if there exist any kind of temporal codification of the olfactory information. Both ICA and MPCA can help in the identification of this stimulatory-inhibitory activity, and become useful tools in this kind of analysis.

Acknowledgements

Authors are grateful to GOSPEL FP6-IST 507610 for its support.

References

Hyvärinen, A., Karhunen, J., Oja, E.: Independent Component Analysis. Wiley, New York (2001)

Kent, P.F., Mozell, M.M.: The recording of odorant-induced mucosal activity patterns with a voltage sensitive dye. J. Neurophys. 68, 1804–1819 (1992)

Orbach, H.S., Cohen, L.B.: Optical monitoring of activity from many areas of the in vitro and in vivo salamander olfactory bulb: a new method for studying functional organization in the vertebrate central nervous system. J. Neurosci. 3, 2251–2262 (1983)

Wise, B.M., Gallagher, N.B., Butler, S.W., White, D.D., Barna, G.G.: A comparison of principal component analysis, Multiway principal component analysis, trilinear decomposition and parallel factor analysis for fault detection in a semiconductor etch process. J. Chemom. 13, 379–396 (1999)

Westerhuis, J.A., Kourti, T., Macgregor, J.F.: Comparing alternative approaches for multivariate statistical analysis of batch process data. J. Chemom. 13, 397–413 (1999)

Shah, M., Persaud, K.C., Polak, E.H., Stussi, E.: Selective and reversible blockage of a fatty acid odour response in the olfactory bulb of the frog (Rana temporaria). Cellular and Molecular Biology 45, 339–345 (1999)

Xueguang, S., Wei, W., Zhenyu, H., Wensheng, C.: A new regression method based on independent component analysis. Talanta 69, 676–680 (2006)

Artificial Olfaction and Gustation

5

Improved Odour Detection through Imposed Biomimetic Temporal Dynamics

Tim C. Pearce[1], Manuel A. Sánchez-Montañés[2], and Julian W. Gardner[3]

[1] Department of Engineering, University of Leicester, Leicester LE1 7RH,
  United Kingdom
[2] Escuela Politécnica Superior, Universidad Autónoma de Madrid, Madrid 28049, Spain
[3] School of Engineering, University of Warwick, Coventry CV4 7AL,
  United Kingdom

Abstract. We discuss a biomimetic approach for improving odour detection in artificial olfactory systems that utilises temporal dynamical delivery of odours to chemical sensor arrays deployed within stationary phase materials. This novel odour analysis technology, which we have termed an artificial mucosa, uses the principle of "nasal chromatography"; thus emulating the action of the mucous coating the olfactory epithelium. Temporal segregation of odorants due to selective phase partitioning during delivery in turn gives rise to complex spatio-temporal dynamics in the responses of the sensor array population, which we have exploited for enhanced detection performance. We consider the challenge of extracting stimulus-specific information from such responses, which requires specialised time-dependent signal processing, information measures and classification techniques.

5.1 Three Key Mechanisms for Discrimination of Complex Odours in Chemical Sensor Arrays

The detection capability of chemical sensor array systems is limited by both sensor noise and the degree to which response properties can be made stimulus specific and diverse across the array (Pearce & Sánchez-Montañés 2003). Two main mechanisms for odour discrimination in artificial olfactory systems have been exploited so far:

1. To generate diverse responses, sensors within the array are typically selected to produce an ensemble of complementary wide spectrum broad tunings to the different volatile compounds of interest. Given sufficient diversity in these tunings, a spatial fingerprint of a particular complex odour should be generated across the array that is sufficiently stimulus specific to overcome noise limitations, and may then be used as part of a pattern recognition scheme for odour discrimination (Pearce *et al.* 2003). In this case, the role of time is not considered explicitly, but rather the magnitudes of the responses across the array, which is the classical method of odour classification in artificial olfactory systems.

A. Gutiérrez and S. Marco (Eds.): Biologically Inspired Signal Processing, SCI 188, pp. 75–91.
springerlink.com   © Springer-Verlag Berlin Heidelberg 2009

2. Depending upon the choice of chemosensor technology and the compounds under investigation, it is possible that the chemosensor dynamics themselves can also depend on the compounds present in the mixture (Albert *et al.* 2002), leading to an additional dimension of temporal variation which can be exploited for the purposes of discrimination. Such differences have previously been exploited to improve discrimination performance (e.g. Llobet *et al.* 1997, White and Kauer 1999).

In biological olfaction, on the other hand, the temporal dimension is known to play a much more central role in the processing of olfactory information (Schoenfeld and Cleland 2006) than has thus far been considered in machine olfaction research. For instance, the timing and dynamics of the sniffing process are known to be important (Kepecs *et al.* 2006), which appears to be well matched with the timing of neural processing mechanisms in the olfactory bulb, as emphasised by many modellers (*e.g.* Brody and Hopfield, 2003). Looking at the overall processes involved in olfactory perception, this may be viewed as an exquisitely timed and orchestrated sequence of odorant inhalation, odorant partitioning and absorption, olfactory neuron timing responses mediated by calcium dynamics, the arrival and complex integration of spikes at glomeruli and the finely balanced dynamics of excitation and lateral inhibition in the bulb. When building biomimetic olfactory systems, therefore, we should consider carefully the timing and temporal aspects of the delivery and processing of sensory information.

By considering the role of timing of odorant delivery in biological olfaction (Rubin & Cleland 2006), we have recently built a novel machine olfaction technology, termed an "artificial olfactory mucosa", which demonstrates clearly a third principle of odour discrimination in artificial olfactory systems:

3. By creating a temporal profile of odour delivery to the different sensors within the array that is stimulus specific, we may provide additional response diversity. This is achieved by deploying chemical sensor arrays within stationary phase materials that impose the necessary stimulus-dependent spatio-temporal dynamics in sensor response; we have recently shown that this approach aids complex odour discrimination (Gardner *et al.* 2007). This concept is very different to that embodied within classical electronic nose systems that are usually designed to control the exposure of the stimulus as a square pulse, whose temporal properties are independent from the nature and chemical composition of the stimulus. Instead, we exploit such differences to generate additional discrimination capability in the device.

5.2   An Artificial Olfactory Mucosa for Enhanced Complex Odour Analysis

This third discrimination mechanism uses the physical positioning of a series of broadly tuned sensors along the length of a planar chromatographic channel (analogous to the thin mucous coating of the nasal cavity) which gives rise to more diversity in the temporal properties in the sensor signals (retentive delay and profile). Figure 5.1 shows the basic architecture of the artificial mucosa concept and its biological counterpart. A complex odour pulse travelling in the mobile carrier phase

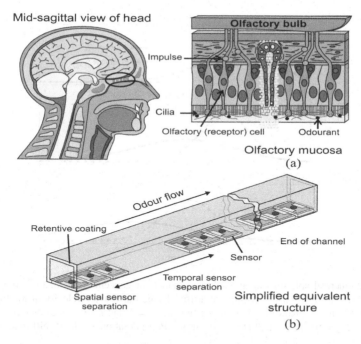

Fig. 5.1. a) Sagittal head view showing the main sections of the olfactory mucosa and subsequent neural processing. Odour molecules during inhalation selectively partition into a mucous layer covering specialized dendritic cilia from olfactory receptor neurons in the nasal epithelium. Odours interact with receptor proteins embedded within the cilia membrane to mediate ORN calcium dynamics, ultimately leading to the generation of additional action potentials (*impulses*). These action potentials are transmitted to the olfactory bulb via axonal projections where these are processed to interpret complex odour information. **b)** An artificial mucosa that relies on similar principles of odorant partitioning to its biological counterpart. The chemosensor array is deployed inside a microchannel coated with a stationary phase material (*retentive coating*) that has selective affinity to the different compounds with a complex mixture. By introducing a pressure difference across the microchannel odour flow may be pulsed within the microchannel, giving rise to segregation in odour components that is compound specific. (Reprinted with permission by Royal Society, London).

inside the artificial mucosa gives rise to selective partitioning of components causing the odour components to travel at different speeds into the mucosa, leading to a kind of chromatographic effect. Depending upon the degree of affinity of each component compound for the retentive layer, this will be found within the mobile (carrier) and stationary (retentive layer) phases in compound specific ratios. The retention of each odour component in the stationary phase acts to retard the progress of the pulse for that compound through the mucosa, leading to segregation in the components of the stimulus in accord with the well understood principles of gas capillary column chromatography (Purnell 1962).

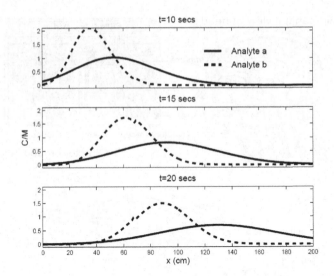

Fig. 5.2. Numerical solutions of an analytical model for concentration for two compounds a and b at different positions, x, within the artificial mucosa. Carrier velocity inside the micro-channel is 15 cm s^{-1}, mass distribution coefficients $k_a = 1$ and $k_b = 2$, and effective diffusion coefficients $D'_a = 50$ cm^2 s^{-1} and $D'_b = 10$ cm^2 s^{-1}. Pulse duration at inlet ("sniff time"), $t = 5$ s.

We will see that this arrangement provides an important additional mechanism for odour discrimination, since depending upon their location in the mucosa, each sensor will receive a particular sequence of single or subsets of compounds within a complex mixture over time, which is a function of the stimulus composition. It is important to understand and accurately describe the transportation of odour compounds within the artificial mucosa in order to verify the experimental results, as well as provide the basis for an optimisation procedure of its design for complex spatiotemporal chemical sensing. We have developed both finite element and analytical models for this purpose. Figure 5.2, for example, shows the numerical solution of our analytical model of local concentration profiles within the micro-channel for two compounds, a and b, injected simultaneously at the inlet as they progress through the device. We see a clear separation between the two compounds that increases over time and depends directly upon the difference in partition coefficients and so is compound and stimulus specific. In both cases the dispersion, which determines the degree of overlap, depends upon the effective diffusion coefficient while the velocity of propagation through the mucosa depends upon the effective partition coefficient between the compound and the stationary phase deployed.

Differential sorption of compounds within the artificial mucosa gives rise to a temporal fingerprint in the chemosensor response which is sensitive to the concentrations and presence of different compounds. The important aspect here that is distinct from previous techniques exploiting the temporal dimension is that the delivery of the stimulus itself becomes specific to the compound(s) being delivered, which imposes additional diversity in the array responses. We have shown experimentally (Gardner *et al.* 2007) that deploying chemical sensor arrays within stationary phase materials in this way imposes stimulus-dependent spatio-temporal dynamics on their response, thereby aiding complex odour discrimination. We will also show theoretically at the end of this chapter that using both spatio-temporal responses (all three discrimination mechanisms) will always provide better detection performance than using spatial information alone (the first discrimination mechanism).

5.2.1 Artificial Olfactory Mucosa Fabrication

The artificial mucosa was constructed by mounting discrete polymer/carbon black composite chemoresistive sensors (40 devices of 10 different composites) on a printed circuit board (PCB) base sealed with two different polyester lids (with and without stationary phase coating, which we refer to here as the coated and uncoated mucosa) within which a serpentine microchannel was machined. Once sealed, this composite structure was injected with Parylene C ,as the absorbent stationary phase material, deposited using a commercial evaporation technique (PDS 2010 Labcoater™ 2, Specialty Coating Systems, Indianapolis, USA). Each sensor chip was 2.5 mm × 4.0 mm in size and comprised a pair of thin co-planar gold electrodes on a SiO_2/Si substrate with an electrode length of 1.0 mm and an inter-electrode gap of 75 μm. Additional fabrication details are provided elsewhere (Gardner *et al.* 2007).

5.2.2 Chemical Sensor Behaviour within the Artificial Mucosa

In order to demonstrate the effect of the stationary phase material within the artificial olfactory mucosa, we tested rectangular pulses of simple odorants (toluene and ethanol) with the microchannel both coated and uncoated – Figure 5.3 shows the normalised results. In both cases, the sensor closest to the inlet of the microchannel (S1) shows a rapid onset time relative to that seen at the sensor towards the outlet (S39), which is due to the transport time for the odour pulse ("sniff"). However, in the uncoated case (Figure 5.3a), we see that the temporal response of the outlet sensor is not stimulus specific in time for ethanol and toluene after normalization. Thus, the uncoated mucosa adds no additional information in time, since within the limits of sensor noise, the outlet sensor is only able to discriminate between the two simple compounds based upon its response magnitudes – *i.e.* using the first mechanism of discrimination. Of particular note here is the broadening of the response signal in time with increasing sensor distance from the inlet, which is also observed in the responses of identical sensors placed at different locations along the channel. This is predominantly due to diffusion broadening of the odour as it travels along the micro-channel. Since the diffusion coefficient in air varies very little for different odour ligands, diffusion broadening, in itself, is not a particularly effective means of imposing stimulus dependent response diversity in artificial olfactory systems. We will see that selective partitioning can play a much more important role.

UNCOATED   COATED

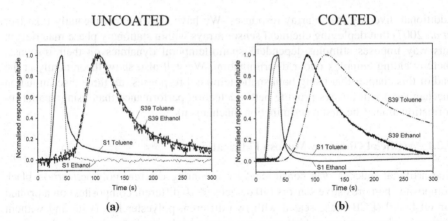

(a)   **(b)**

Fig. 5.3. Comparison of normalized chemosensor responses for an uncoated and coated artificial olfactory mucosa. **a)** Uncoated mucosa. Responses of sensor S1 (PEVA sensor material composite) close to the inlet and S39 (PCL sensor material composite) close to the outlet of the microchannel. **b)** Responses from the same sensors in the coated mucosa. (Reprinted with permission by Royal Society, London).

The uncoated responses of the inlet sensor also show some differential response that is stimulus dependent, which is most likely due to the kinetics of the ligand/sensor interaction rather than the mucosa properties – an example of the second discrimination mechanism discussed in Section 1 due to differential ligand/sensor temporal interactions.

In the coated case (the normal operational mode of the artificial mucosa - Figure 5.3b), the response of the outlet sensor after normalization shows very different temporal responses that are strongly stimulus specific. Here we see a clear additional latency in the onset of the response and also its duration is much longer, which is clearly due to spatio-temporal stimulus dynamics imposed by the coated mucosa when we compare to the uncoated case. This stimulus dependent difference in

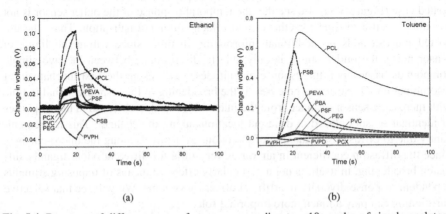

(a)   (b)

Fig. 5.4. Response of different types of sensors responding to a 10 s pulse of simple analytes. Sensor responses to **a)** ethanol vapour, and **b)** toluene vapour in air. (Reprinted with permission by Royal Society, London).

the outlet sensor response, demonstrates clearly the additional third mechanism for discrimination which we have produced through the use of selective coatings in our artificial mucosa design and believe to be analogous to odorant air/mucus interaction in the biological olfactory system. Figure 5.4 shows the diversity in the sensor responses for the different composite materials we have used within the artificial mucosa device. The ensemble response clearly shows a wide diversity that is strongly stimulus specific, both before and after normalisation. This additional temporal diversity due to selective partitioning is a powerful means for introducing additional discrimination capability to chemical sensor arrays for complex odour analysis.

5.3 Exploiting Temporal Responses in the Artificial Mucosa

The three mechanisms for discrimination in artificial olfactory systems are not mutually exclusive. Rather, as appears to be the case in biology, we may exploit differential responses due to diverse sensor tunings (first mechanism), ligand/sensor kinetic dependencies (second mechanism), and imposed spatio-temporal dynamics in the stimulus delivery (third mechanism) simultaneously. Making these three mechanisms for discrimination cooperate in tandem and in a selective way is, we believe, the key to building new generations of artificial olfactory systems that may begin to approach the impressive selectivity and broad range sensitivity found in biological systems.

In order to take advantage of the rich diversity of temporal responses created by the artificial mucosa we must analyse them with suitable signal processing and classification strategies, i.e. techniques that are time dependent. One approach is to again look to the biology for the principles involved in processing such spatio-temporal signals (Pearce 1997).

5.3.1 Olfactory Bulb Implementations for Spatiotemporal Processing of Odour Information

A large number of olfactory receptor neurons (ORNs) constitute the front-end of the olfactory system, being responsible for detecting airborne molecules. Cilia of the ORNs protrude into the olfactory mucosa (Figure 5.1), where they come in contact with molecules that are transported by the nasal air flow. On the surface of the cilia, odorant receptors bind odorant molecules with a broadly tuned affinity. When a receptor binds with an odorant molecule, it triggers in its ORN a biochemical cascade that eventually causes the membrane potential of the ORN to change, potentially leading to the generation of action potentials (Mori *et al.* 1999).

In vertebrates, ORNs project their axons into the olfactory bulb, terminating into spherical neuropils called glomeruli, where they connect onto the dendrites of mitral and tufted (M/T) cells. Experimental data indicate that each glomerulus receives the axons of only ORNs that express the same type of receptor, while any M/T cell sends its apical dendrite into one glomerulus only. Inhibitory neurons of the olfactory bulb form reciprocal contacts with many M/T cells via granule cells, thus forming together a complex network that appears to constitute the first stage of olfactory information processing. The output of the M/T cells is also relayed to higher brain areas (Mori *et al.* 1999).

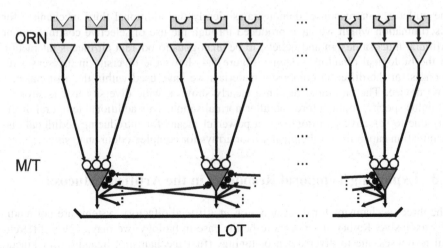

Fig. 5.5. A schematic diagram of the olfactory bulb neuronal model architecture which we have implemented in programmable logic (Guerrero and Pearce 2007) and aVLSI (Koickal *et al.* 2007) for real-time odour signal processing, showing receptor and principal neurons (triangles) and synapses (circles: unfilled – excitatory, filled – inhibitory). There are 25 M/T cells in total and 75 ORNs. M/T: mitral/tufted cells, ORN: olfactory receptor neurons. LOT: lateral olfactory tract.

Figure 5.5 shows the overall schematic of our network model for spatio-temporal odour signal processing (only showing three types of receptor input for the sake of clarity). The diagram has been drawn representing every computational element with an individual device, rather than adopting a biologically realistic representation. Chemosensors themselves are represented by irregular polygons at the top of the diagram which may be placed within the artificial mucosa to generate additional temporal diversity in their responses – polygons of the same shape represent sensors of the same type. Since the firing rate produced by ORNs is limited to approx. 1 kHz, we use a sigmoidal squashing function to condition the chemosensor signals before using it to drive the olfactory bulb (OB) model.

In our model, any ORN receives input from only one chemosensor/receptor type, and any chemosensor only projects to one ORN. The outputs of the ORNs feed into the respective ORN-M/T synapses (circles). The outputs of the synapses that receive input from ORNs converge a single M/T cell, where they are summed linearly. This represents the operation of glomeruli in the olfactory system. The output of any glomerulus feeds into one respective M/T cell. Because the signals from sensors of the same type are fed forward through neural elements to a single M/T cell, the network presents an evident modular structure, each module being defined by a different type of sensor, in a way that resembles the glomerular organization of the olfactory bulb. Every M/T cell projects to every other M/T cell through one of the M/T-M/T inhibitory synapses (filled circles).

The neurons themselves have been modelled as integrate-and-fire units. Below the threshold V_θ, the dynamics of the "membrane" potential $V(t)$ of the IF neuron are defined by the equation

$$\frac{dV(t)}{dt} = -\frac{V(t)}{RC} + \frac{I(t)}{C} \tag{5.1}$$

where t is time, R and C are, respectively, the membrane resistance and capacitance, and $I(t)$ is the total input current to the neuron. The membrane rest potential is conventionally set equal to the value zero. The membrane time-constant τ_m is given by RC ($\tau_m = 10$ ms is used throughout as a biologically plausible value). The terms contributing to $I(t)$ are due to sensor responses if the neuron is a ORN, and to ORNs and lateral interactions if the neuron is a M/T cell. If the potential $V(t)$ reaches the threshold value V_θ, it is immediately reset to the afterhyperpolarisation value V_{ahp} and an action potential is produced as output of the neuron. Since we do not explicitly consider the role of adaptation in the model at this time, we set the threshold V_θ to be equal and fixed for all ORN and M/T cells.

The model also includes dynamical synapses based upon first order dynamics. In this case, currents generated by a synapse in response to a spike train is through an exponential decay over multiple spike inputs occurring at times $(t_1, t_2, \ldots, t_j, \ldots t_l)$ to give the dendritic current

$$I(t) = w \sum_{j=1}^{l} H(t - t_j) \exp\left\{-\frac{t - t_j}{\tau_{i;e}}\right\}, \tag{5.2}$$

where w specifies the weight or efficacy, $H(.)$ is the Heaviside function and $\tau_{i;e}$ is the synaptic time constant of either inhibitory or excitatory synapses. We choose a value of 4 ms for excitatory synapses, and 16 ms for GABA mediated inhibitory synapses.

The weights of the M/T-M/T synapses are defined so as to endow the network with associative properties (cf., e.g., Amit 1992 for general definitions), although it is not known whether the biological counterpart does perform this kind of processing. This lateral connectitivity represents dendrodendritic mitral/tufted and granule cell interactions in the external plexiform layer of the bulb, which are known to be important in mediateing odour memory (Hildebrand and Shepherd, 1997). The network learns odorants by modifying the weights of the M/T-M/T synapses according to a Hebbian learning rule, during a training stage. The activity of a given neuron i by means of its firing rate v_I (defined by a temporal average of spikes, hence, the mean firing rate v is the number of spikes $n_{sp}(t)$ that occur in the time T, $v = n_{sp}(T) / T$, where the v is expressed in Hz). The synaptic weight change is then given by the equation

$$\delta w_{ij} = \alpha \cdot v_i \cdot v_j, \tag{5.3}$$

where v_i and v_j are the firing rates of the postsynaptic and presynaptic cells respectively, and α is a learning parameter, such that $\alpha > 0$. Since the spatial dependence of granule-mitral cell interactions is not fully understood, we choose the lateral weights, w_{ij} to be random before training.

Given a learnt Odour A, an indicator function of all M/T cell firing rates can be defined

$$f = k\frac{\mathbf{d}^T\mathbf{a}}{|\mathbf{d}||\mathbf{a}|}, \tag{5.4}$$

that indicates to which degree the network output is currently representing that particular odorant. Here **d** and **a** are vectors of the M/T cell population firing rates after training in response to the currently presented odour D and previously presented target Odour A. The indicator function is normalised by the magnitude of these vectors so that this may be interpreted as a correlation between **d** and **a**, scaled by k, which is kept constant for all indicator functions. Thus, when a previously learnt odorant is presented, its corresponding indicator function should assume a relatively large positive value.

We have implemented this spatio-temporal olfactory bulb network in both programmable logic (Guerrero & Pearce, 2007, Guerrero-Rivera et al., 2006) and in analogue VLSI technology (Koickal *et al.*, 2007) for the purposes of real-time processing for artificial mucosa. Such recurrent spiking neuronal models have been shown to exhibit Hopfield-like attractor based dynamical behaviour. The asymmetric nature of the connectivity in our model gives rise to a richer variety of dynamical behaviours than in the symmetric Hopfield case (Li and Dayan, 1999). Such networks have been shown in a variety of contexts to be sensitive to temporal properties in their input, for instance, temporal sequence processing (Wang 2003). We next show that the network is capable of supporting odour classification and odour compound detection in varying backgrounds.

A. Odour Classification

In order to demonstrate the classification properties of the OB model, the same network was subjected to two arbitrary but constant input patterns (that we term 'Odour A' and 'Odour B'), representing the receptor response to distinct odours at the input.

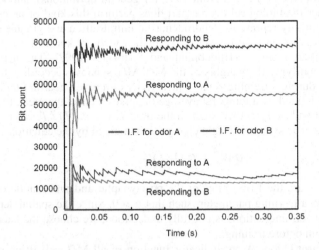

Fig. 5.6. Indicator functions for the odour classification task. The same network is trained to 2 odours, 'A' and 'B' from which indicator functions (I.F.) are constructed. Shown is the indicator function response for both A and B when odour A and B are presented. In each case the indicator function for the learnt odour is far higher than that for the distractor odour.

Hebbian learning was first applied to adapt the lateral weights of the network when exposed to Odour A and then applied a second time for 'Odour B'. After training, each odour ('Odour A' and 'Odour B') was presented sequentially for testing and the corresponding indicator functions for both learnt odours calculated over time (see Figure 5.6). We see that the corresponding indicator function for the learnt odour is high after a short time, whereas the other indicator function is low. When we present the second odour the situation is the opposite, indicating that the network is able to store two attractors corresponding to the two odours and we can read these out to classify the odours accordingly.

B. Odour Identification in Interfering Backgrounds

In order to test how robust the learnt odour classification scheme is in this case, we trained the system to a random pattern of steady-state activity across the array, which we termed 'Odour A' and used the network response to define the indicator function. When this input was presented, the pure Odour A stimulus gave rise to a large indicator function response, shown in Figure 5.7. In order to confound the input pattern, we then added different fractions of a random pattern, 'Odour B' to the original odour. In this case, the indicator function was found to reliably identify the presence of 'Odour A' even when the original learnt odour response was linearly superimposed on the distractor odour. To be sure that such a large indicator function response did not occur by chance or across all possible stimuli, Figure 5.7 also shows a low indicator function response in the trained network to a 3rd random odour input presented separately, 'Odour C'. Our demonstration of the ability of the network to solve this task corresponds to an important problem of identifying a learnt odour when presented in the context of some unknown, distractor, chemical such as in an explosives detection task.

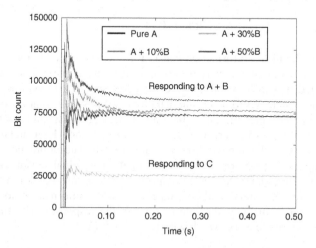

Fig. 5.7. Indicator functions for the odour identification in interfering background task. The network is trained to odour 'A' from which indicator function (I.F.) is constructed. Shown is the indicator function response for pure odour A and odour A mixed with various concentrations of a distractor odour B. Additionally, an odour C is applied to demonstrate low activity to untrained odours.

Such biologically-plausible networks will further be considered for their properties in complex odour detection tasks that have thus far not been solved using classical signal processing and pattern recognition approaches.

5.3.2 Spatiotemporal Information Measures

Fisher Information

In previous work we have discussed and analyzed how Fisher Information can be used to quantify the performance of an electronic nose (Sánchez-Montañés and Pearce 2001, Pearce and Sánchez-Montañés 2003). Basically, the *Fisher Information Matrix* (FIM) **F** is a square and symmetric matrix of $s \times s$ components, where s is the number of individual compounds whose concentration we are interested to estimate. In order to calculate **F** we should first calculate the individual FIMs for each sensor j:

$$\mathbf{F}_j = \int p(\mathbf{Y}_j \mid \mathbf{c}) \cdot \left[\frac{\partial p(\mathbf{Y}_j \mid \mathbf{c})}{\partial \mathbf{c}} \right] \cdot \left[\frac{\partial p(\mathbf{Y}_j \mid \mathbf{c})}{\partial \mathbf{c}} \right]^T \cdot d^L \mathbf{Y}_j \tag{5.5}$$

where $\mathbf{Y}_j$ is the response of sensor j and $\mathbf{c}$ is a vector with the concentrations of the s simple compounds. The equation is general in that the sensor response $\mathbf{Y}$ may either be a time-independent scalar or a time series vector (of dimension L). In this case the total FIM for the array is just the summation of the individual matrices for each sensor, $\mathbf{F}_j$. The probability distribution $p(\mathbf{Y}_j|\mathbf{c})$ represents the noisy response of the sensor to a given mixture with concentration vector $\mathbf{c}$ (odour space).

The usefulness of Fisher Information is given by the important property that the best square error across all unbiased techniques that use the noisy array responses to estimate the stimulus is (see Sánchez-Montañés and Pearce 2001 for discussion)

$$\varepsilon^2 = \text{trace} (\mathbf{F}^{-1}) \tag{5.6}$$

Importantly, the Fisher Information matrix **F** is closely related to the discrimination ability of the system, which is why we consider it in this context. For instance, it can be demonstrated that in a two-alternative forced choice discrimination between two stimuli (i.e. the system has to determine which of two possible complex odours $\mathbf{c}_1$ and $\mathbf{c}_2$ is being presented), the optimal probability of error $P(\epsilon)$ that can be achieved using linear sensors is $P(\epsilon) = 0.5 \cdot [1 - \text{erf}(0.5 \cdot \lambda^{0.5})]$ with $\lambda \equiv \frac{1}{2} \cdot \delta \mathbf{c}^T \cdot \mathbf{F} \cdot \delta \mathbf{c}$ and $\delta \mathbf{c} \equiv \mathbf{c}_2 - \mathbf{c}_1$.

In our previous work we have discussed how to calculate in practice this quantity when the temporal patterns of the responses of the individual sensors are not taken into account (corresponding to the first mechanism for discrimination identified in Section 1). Here we extend and calculate the Fisher Information for the spatio-temporal case which includes the role of time in the responses. The first step is to model the noise in the sensors, which will determine $p (\mathbf{Y}_j \mid \mathbf{c})$.

Dynamic Model of the Noise

Let us define $\mathbf{Y}_{j;\, \mathbf{c}}$ as the noisy temporal response (time series) of sensor j to stimulus $\mathbf{c}$. $\mathbf{Y}_{j;\, \mathbf{c}}$ is a vector of L components (number of consecutive samples of the sensor). We will consider sensors with additive noise,

$$\mathbf{Y}_{j;\mathbf{c}} = \overline{\mathbf{Y}}_{j;\mathbf{c}} + \mathbf{n}_j \tag{5.7}$$

where $\overline{\mathbf{Y}}_{j;\mathbf{c}}$ is the expected time series response of sensor j to mixture $\mathbf{c}$, and $\mathbf{n}_j$ is a noisy time series that corrupts the individual sensor response. Note that $\overline{\mathbf{Y}}_{j;\mathbf{c}}$ can be in principle a time series of arbitrary complexity, for instance a series with four different peaks. In order to calculate the Fisher Information we need to characterize the noise dynamics of $\mathbf{n}_j$. To first approximation, we model them as first-order AR processes, which we express in the convenient form

$$\mathbf{n}_j(k+1) = \gamma_j \cdot \mathbf{n}_j(k) + \sigma_j \cdot \sqrt{1-\gamma_j^2} \cdot \xi \tag{5.8}$$

with $k \in [1, L\text{-}1]$. Additionally, $\mathbf{n}_j(1)$ is modelled as a Gaussian random variable of zero mean and variance σ_j^2. In equation 6 and from now on we use parentheses to indicate the element of the vector (k) or matrix (u, v) to avoid confusion with subscripts. The coefficients γ_j and σ_j depend on each sensor; ξ is an I.I.D. Gaussian variable of unit variance and zero mean. Note that this implies that $\mathbf{n}_j$ has zero mean, variance σ_j^2 and auto-covariance given by:

$$\left\langle \mathbf{n}_j(k) \cdot \mathbf{n}_j(k+d) \right\rangle = \sigma_j^2 \cdot \gamma_j^{|d|} \tag{5.9}$$

Therefore the noise vector $\mathbf{n}_j$ is a multivariate Gaussian process with zero mean and covariance matrix $\mathbf{N}_j$ given by

$$\mathbf{N}_j(u,v) = \sigma_j^2 \cdot \gamma_j^{|u-v|} \tag{5.10}$$

where $u, v \in 1,...,L$

Spatio-temporal Fisher Information

The expected time series response of a linear sensor j to a mixture $\mathbf{c}$ is given by (removing the constant sensor baseline):

$$\overline{\mathbf{Y}}_{j;\mathbf{c}} = \Sigma_i \, c_i \cdot \mathbf{A}_j^i \tag{5.11}$$

where $\mathbf{A}_j^i$ is the expected time series response of sensor j to a unit of concentration of single compound i. Equation 9 implies that the sensor response is linear to increasing concentration (within some reasonable limit) and to mixtures. We have found that this is a good approximation for the composite materials used in our artificial mucosa (data not shown).

Using the previous result that $\mathbf{Y}_{j;\mathbf{c}}$ is a Gaussian vector with covariance matrix $\mathbf{N}_j$, together with equation 3, it is straightforward to demonstrate

$$\mathbf{F}_j(u,v) = \left(\mathbf{A}_j^u\right) \cdot \mathbf{N}_j^{-1} \cdot \left(\mathbf{A}_j^v\right)^T \tag{5.12}$$

Using this equation together with equation 8 and using that sensors are causal ($\mathbf{A}_j$ $(1, k) = 0$) we can derive after some algebra the following convenient equation

$$\mathbf{F}_j(u,v) = \frac{1}{\sigma_j^2 \cdot \left(1 - \gamma_j^2\right)} \cdot \sum_{i=2}^{L} \mathbf{B}_j(i,u) \cdot \mathbf{B}_j(i,v) \qquad (5.13)$$

with $\mathbf{B}_j(i, k) \equiv \mathbf{A}_j(i, k) - \gamma_j \cdot \mathbf{A}_j(i\text{-}1, k)$. This important equation represents the spatio-temporal Fisher Information of a noisy sensor within an array.

Purely Spatial Fisher Information

It is interesting to calculate how much better the spatio-temporal information is when compared to the information carried by sensor responses containing no explicit temporal information. Here we consider the contribution of the three mechanisms combined. In case that just the mean output of each sensor is used in subsequent signal processing:

$$y_{j;\mathbf{c}} = \frac{1}{L} \cdot \sum_{i=1}^{L} \mathbf{Y}_{j;\mathbf{c}}(i) \qquad (5.14)$$

it is easy to demonstrate that this mean output is a Gaussian variable with average $\mathbf{a}_j^T \cdot \mathbf{c}$, where

$$a_j(u) = \frac{1}{L} \cdot \sum_{i=1}^{L} \mathbf{A}_j(i,u) \qquad (5.15)$$

The variance of $y_{j;\mathbf{c}}$ can be calculated as:

$$\sigma_j^2\left(y_{j;\mathbf{c}}\right) = \frac{\sigma_j^2}{\left(1 - \gamma_j\right)^2 \cdot L^2} \cdot \left[L - \gamma_j^2 L - 2\gamma_j\left(1 - \gamma_j\right)\right] \qquad (5.16)$$

Then the individual Fisher Information Matrices are given by (Sánchez-Montañés and Pearce 2001):

$$\mathbf{F}_j = \frac{1}{\sigma_j^2\left(y_{j;\mathbf{c}}\right)} \cdot \mathbf{a}_j \cdot \mathbf{a}_j^T \qquad (5.17)$$

Fisher Information of the Microchannel to Mixtures of Toluene and Ethanol

The spatio-temporal Fisher Information Matrix was calculated for the micro-channel responding to toluene and ethanol odorants, such as that shown in Figure 5.3. Then the trace of the inverse of this matrix was calculated which corresponds to the optimal square error that any method estimating the individual concentrations could obtain.

Importantly, we see that the expected square error when using spatio-temporal information is always smaller than that error when only spatial information is considered (Figure 5.8). When using all the 16 available sensors in the array the ratio of the

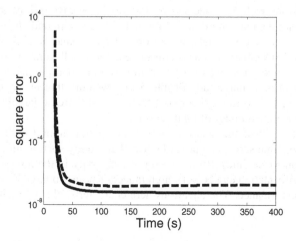

Fig. 5.8. Optimal square error in the estimation of the individual concentrations in mixtures of toluene and ethanol, as a function of the sampling time. All available sensors in the array are used. **Solid:** optimal square error $\epsilon^2_{\text{sp-temp}}$ when the spatial and temporal information in the sensor array is taken into account. **Dashed:** optimal square error ϵ^2_{sp} when only the spatial information is taken into account.

two square errors is 2.8. When the detection task is much harder, *i.e.* complex odours with large numbers of components, this ratio of the spatial error to spatio-temporal error is expected to be far higher.

In order to investigate more deeply how much improved the spatio-temporal information is with respect to pure spatial information (for the case of arbitrary mixtures of the two pure odours toluene and ethanol), we calculated the ratio of the two optimal estimation errors $\epsilon^2_{\text{sp-temp}}$ and ϵ^2_{sp} for all the 65,535 sets of sensors that can be generated out of our 16 available sensors (Table 5.1). Specifically, for each possible subset of sensors we have performed analogous calculations as those shown in Figure 5.8, and then computed the ratio of the minima of the two curves. Table 5.1 shows the resulting range of ratios for a given number of sensors.

Table 5.1. Range of the ratio $\epsilon^2_{\text{sp-temp}} : \epsilon^2_{\text{sp}}$ for all the possible combinations that can be generated out of our 16 available sensors

No. sensors	Improvement	No. sensors	Improvement
1	∞	9	1.6 – 50
2	1.5 – 860	10	1.7 – 33
3	1.4 – 580	11	1.7 – 18
4	1.4 – 370	12	1.8 – 5.8
5	1.5 – 270	13	2.1 – 3.9
6	1.5 – 170	14	2.2 – 3.5
7	1.5 – 83	15	2.3 – 3.1
8	1.6 – 72	16	2.8

For configurations with only one sensor, the ratio is always infinite since it is impossible to estimate the concentrations of the individual compounds from the average response of only one sensor, a task which on the other hand is possible to address when the temporal information is also considered. For small numbers of sensors the improvement of the performance of the system based on spatio-temporal information can be several orders of magnitude (Table 5.1), revealing that an artificial mucosa optimally designed for exploiting temporal features can increase largely the sensitivity as well as reduce the number of total sensors.

The important result of this analysis is that the spatiotemporal information available from a sensor time series can never be less than purely spatial information (such as the mean output over time). By increasing the diversity in the temporal responses the ratio of this information can be very high indeed, leading to large improvements in discrimination performance. The equations here characterise this explicitly in the linear case.

5.4 Conclusions

Achieving optimal detection performance in machine olfaction means exploiting both spatial and temporal sensor array responses, whereas traditionally only the spatial aspects have been employed. We have presented a new machine olfaction technology that demonstrates an additional mechanism for discrimination in these systems, which we have termed an artificial olfactory mucosa, on account of its similarities to biological odour delivery systems. The additional discrimination mechanism acts through the physical segregation in complex mixtures of odours combined with chemical sensor arrays that are distributed in space. Imposing spatio-temporal dynamics in the delivery of chemical components, we have shown, can confer additional diversity in the responses of chemosensor arrays which may form the basis of a new generation of electronic noses with improved sensitivity, discrimination performance and selectivity.

Taking advantage of this new sensing approach requires the consideration of both space and time during chemosensor array signal processing and classification. Here we have emphasised how a spiking implementation of the olfactory bulb, which is also biologically plausible, is able to learn and classify different olfactory inputs as well as identify particular odour stimuli present within a mixture of interfering distractor odorants.

More formally, a new information theory measure has been described which is capable of quantifying both spatial and temporal information in artificial mucosa based chemical sensor arrays. Importantly this analysis has demonstrated that the spatio-temporal case should outperform the purely spatial case emphasising the importance of time in these systems.

The artificial mucosa arrangement opens various possibilities for optimising both spatial and temporal response profiles to particular compounds and mixtures of interest – for instance by configuring sensor position. We are now applying this new information measure to the optimisation of artificial mucosa configurations to particular detection tasks which will uncover underlying design principles for making a new generation of complex odour detection devices with improved detection capabilities.

References

Albert, K.A., Gill, D.S., Walt, D.R., Pearce, T.C.: Optical Multi-bead Arrays for Simple and Complex Odour Discrimination. Analytical Chemistry 73, 2501–2508 (2001)

Amit, D.J.: Modeling Brain Function: The World of Attractor Neural Networks. Cambridge University Press, Cambridge (1992)

Brody, C., Hopfield, J.J.: Simple networks for spike timing computation and olfaction. Neuron. 37, 843–852 (2003)

Hildebrand, J.G., Shepherd, G.M.: Mechanisms of olfactory discrimination: converging evidence for common principles across phyla. Annu. Rev. Neurosci. 20, 595–631 (1997)

Gardner, J.W., Covington, J.A., Koickal, T.J., Hamilton, A., Tan, S.L., Pearce, T.C.: Towards an artificial human olfactory mucosa for improved odour classification. Proceedings of the Royal Society A: Mathematical, Physical and Engineering Sciences 463, 1713–1728 (2007)

Guerrero, R., Pearce, T.C.: Attractor-Based Pattern Classification in a Spiking FPGA Implementation of the Olfactory Bulb. In: Proceedings of the 3rd International IEEE EMBS Conference on Neural Engineering, Hawaii, USA (May 2007)

Guerrero-Rivera, R., Morrison, A., Diesmann, M., Pearce, T.C.: Programmable Logic Construction Kits for Hyper Real-time Neuronal Modeling. Neural Computation 18, 2651–2679 (2006)

Kepecs, A., Uchida, N., Mainen, Z.F.: The sniff as a unit of olfactory processing. Chemical Senses 31, 167–179 (2006)

Koickal, T.J., Hamilton, A., Tan, S.L., Covington, J.A., Gardner, J.W., Pearce, T.C.: Analog VLSI Circuit Implementation of an Adaptive Neuromorphic Olfaction Chip. IEEE Circuits and Systems 54, 60–73 (2007)

Li, Z., Dayan, P.: Computational differences between asymmetrical and symmetrical networks Network. Comput. Neural Syst. 10, 59–77 (1999)

Llobet, E., Brezmes, J., Vilanova, X., Sueiras, J.E., Correig, X.: Qualitative and quantitative analysis of volatile organic compounds using transient and steady-state responses of a thick-film tin oxide gas sensor array. Sensors and Actuators B - Chemical 41, 13–21 (1997)

Mori, K., Nagao, H., Yoshihara, Y.: The olfactory bulb: coding and processing of odour molecule information. Science 286, 711 (1999)

Pearce, T.C.: Computational Parallels between the Biological Olfactory Pathway and its Analogue The Electronic Nose: Part I Biological Olfaction. BioSystems 41, 43–67 (1997)

Pearce, T.C., Gardner, J.W., Nagle, H.T., Schiffman, S.S. (eds.): Handbook of Machine Olfaction. Wiley-VCH, Weinheim (2003)

Pearce, T.C., Sánchez-Montañés, M.A.: Chemical Sensor Array Optimization: Geometric and Theoretical Approaches. In: Handbook of Machine Olfaction: Electronic Nose Technology. In: Pearce, T.C., Schiffman, S.S., Nagle, H.T., Gardner, J.W. (eds.) Handbook of Machine Olfaction: Electronic Nose Technology, pp. 347–375. Wiley-VCH (2003)

Purnell, H.: Gas Chromatography. John Wiley & Sons, Chichester (1962)

Rubin, D.B., Cleland, T.A.: Dynamical mechanisms of odour processing in olfactory bulb mitral cells. J. Neurophysiol. 96(2), 555–568 (2006)

Sánchez-Montañés, M.A., Pearce, T.C.: Fisher Information and Optimal Odour Sensors. Neurocomputing 38, 335–341 (2001)

Schoenfeld, T.A., Cleland, T.A.: Anatomical contributions to odorant sampling and representation in rodents: zoning in on sniffing behavior. Chem. Senses 31, 131–144 (2006)

Wang, D.: Temporal pattern processing. In: Arbib, M. (ed.) The Handbook of Brain Theory and Neural Network, vol. 2, pp. 1163–1167. MIT Press, Cambridge (2003)

White, J., Kauer, J.S.: Odour recognition in an artificial nose by spatio-temporal processing using an olfactory neuronal network. Neurocomputing 26(7), 919–924 (1999)

6

Relating Sensor Responses of Odorants to Their Organoleptic Properties by Means of a Biologically-Inspired Model of Receptor Neuron Convergence onto Olfactory Bulb

Baranidharan Raman[1,2] and Ricardo Gutierrez-Osuna[1]

[1] Department of Computer Science, Texas A&M University,
520 Harvey R. Bright Bldg, College Station, TX 77843-3112, USA
[2] Process Sensing Group, Chemical Science and Technology Laboratory,
National Institute of Standards and Technology (NIST), 100 Bureau Drive MS8362,
Gaithersburg, MD 20899-8362, USA
baranidharan.raman@nist.gov, rgutier@cs.tamu.edu

Abstract. We present a neuromorphic approach to study the relationship between the response of a sensor/instrument to odorant molecules and the perceptual characteristics of the odors. Clearly, such correlations are only possible if the sensing instrument captures information about molecular properties (e.g., functional group, carbon chain-length) to which biological receptors have affinity. Given that information about some of these molecular features can be extracted from their infrared absorption spectra, an attractive candidate for this study is infrared (IR) spectroscopy. In our proposed model, high-dimensional IR absorption spectra of analytes are converted into compact, spatial odor maps using a feature clustering scheme that mimics the chemotopic convergence of receptor neurons onto the olfactory bulb. Cluster analysis of the generated IR odor maps reveals chemical groups with members that have similar perceptual characteristics e.g. fruits, nuts, etc. Further, the generated clusters match those obtained from a similar analysis of olfactory bulb odor maps obtained in rats for the same set of chemicals. Our results suggest that convergence mapping combined with IR absorption spectra may be an appropriate method to capture perceptual characteristics of certain classes of odorants.

6.1 Introduction

Smell is the most primitive of the known senses. In humans, smell is often viewed as an aesthetic sense, as a sense capable of eliciting enduring thoughts and memories. For many animal species however, olfaction is the primary sense. Olfactory cues are extensively used for food foraging, trail following, mating, bonding, navigation, and detection of threats (Axel 1995). Irrespective of its purpose i.e., as a primary sense or as an aesthetic sense, there exists an astonishing similarity in the organization of the peripheral olfactory system across phyla (Hildebrand and Shepherd 1997). This suggests that the biological olfactory system may have been optimized over evolutionary time to perform the essential but complex task of recognizing odorants from their molecular features, and generating the perception of smells.

A. Gutiérrez and S. Marco (Eds.): Biologically Inspired Signal Processing, SCI 188, pp. 93–108.
springerlink.com   © Springer-Verlag Berlin Heidelberg 2009

Inspired by biology, artificial systems for chemical sensing and odor measurement, popularly referred to as the 'electronic nose technology' or 'e-noses' for short, have emerged in the past two decades. A number of parallels between biological and artificial olfaction are well known to the e-nose community. Two of these parallels are at the core of sensor-based machine olfaction (SBMO), as stated in the seminal work of Persaud and Dodd (1982). First, biology relies on a population of olfactory receptor neurons (ORNs) that are broadly tuned to odorants. In turn, SBMO employs chemical sensor arrays with highly overlapping selectivities. Second, neural circuitry downstream from the olfactory epithelium improves the signal-to-noise ratio and the specificity of the initial receptor code, enabling wider odor detection ranges than those of individual receptors. Pattern recognition of chemical sensor signals performs similar functions through preprocessing, dimensionality reduction, and classification/regression algorithms.

Most of the current approaches for processing multivariate data from e-noses are direct applications of statistical pattern recognition and chemometrics techniques (Gutierrez-Osuna 2002). In this book chapter, we focus on an alternative approach: a computational model inspired by information processing in the biological olfactory system. This neuromorphic approach to signal processing represents a unique departure from current practices, one that could move us a small step beyond multivariate chemical sensing and in the direction of true machine olfaction: relating sensor/instrumental signals to the perceptual characteristics of the odorant being sensed.

6.2   Odor Representation in the Early Stages of the Olfactory Pathway

The first stage of processing in the olfactory pathway consists of a large array (~10-100 million) of olfactory receptor neurons (ORNs), each of which selectively expresses one or a few genes from a large (100-1,000) family of receptor proteins (Buck and Axel, 1991; Firestein, 2001). Each receptor is capable of detecting multiple odorants, and each odorant can be detected by multiple receptors, leading to a massively combinatorial olfactory code at the receptor level. It has been shown (Alkasab et al. 2002; Zhang and Sejnowski 1999) that this broad tuning of receptors may be an advantageous strategy for sensory systems dealing with a very large detection space. This is certainly the case for the human olfactory system, which has been estimated to discriminate up to 10,000 different odorants (Schiffman and Pearce, 2003). Further, the massively redundant representation improves signal-to-noise ratio, providing increased sensitivity in the subsequent processing layers (Pearce et al. 2002).

Receptor neurons relay their responses downstream to the olfactory bulb (OB) for further processing. Receptor neurons expressing the same receptor gene converge onto one or a few glomeruli (GL) (Mori et al. 1999; Laurent 1999), which are spherical structures of neuropil at the input of the OB. This form of convergence serves two computational functions. First, massive summation of multiple ORN inputs averages out uncorrelated noise, allowing the system to detect odorants below the detection threshold of individual ORNs (Pearce et al., 2002). Second, chemotopic organization leads to a more compact odorant representation, an odor map that encodes odor identity/ quality (Friedrich and Korsching 1997). The generated odor maps have also been shown to correlate with the overall odor percept (Leon and Johnson 2003; Uchida et al., 2000). Hence we will focus on these convergence circuits in this study.

6.3 Infrared Absoption Spectroscopy

Though very little is known about the molecular determinants of an odorant, it is widely believed that each glomerulus (to which similar ORNs converge) acts as a "molecular feature detector" that identifies a particular molecular property, such as type and position of a functional group (Mori et al. 1999) or carbon chain-length (Sachse et al., 1999). Information about these molecular features can be extracted from their IR absorption spectra, making IR absorption spectroscopy an attractive candidate for this study.

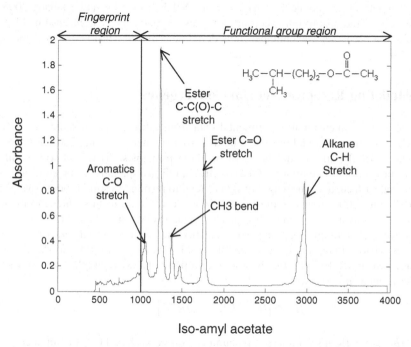

Iso-amyl acetate

Fig. 6.1. IR absorption spectrum of iso-amyl acetate (an ester with a fruity smell). Each peak is labeled by the functional group responsible for the absorption.

Infrared radiations are electromagnetic waves whose wavelength lies in the region between the visible light and microwaves. When exposed to IR rays, molecules tend to absorb these radiations at wavelengths where the radiant energy matches the energy of their intra-molecular vibrations. In IR spectroscopy, differences in molecular structure and inter-atomic bonds between chemicals are exploited to generate unique IR absorption spectra that are rich in analytical information (Nogueira et al. in press). The entire IR spectrum comprises of three non-overlapping regions, each with a distinct purpose: (1) the far-IR region (400-10 cm^{-1}), used for rotational spectroscopy, (2) the mid-IR region (4000-400 cm^{-1}), which provides information about molecular rotations-vibrations, and (3) the near-IR region (12800-4000 cm^{-1}), used for studying molecular overtones and certain combination vibrations. Of particular interest is the mid-IR region, which is further subdivided into the so-called "functional-group" (4000-1500 cm^{-1}) and "fingerprint" (<1500 cm^{-1}) regions. The former contains information about the functional groups that are present in the molecule (e.g., alcohols, aldehydes, ketones, esters etc.,), whereas the latter contains a global absorption pattern that is unique to each organic compound. A sample IR spectrum (iso-amyl acetate; an ester with a fruity smell) obtained from the National Institute of Standards and Technology (NIST) Chemistry Web Book database (Linstrom and Mallard 2003) is shown in Figure 6.1. Different peaks in the absorption spectrum correspond to the various molecular features present in iso-amyl acetate.

We use a database comprising of IR absorption spectra (wave number range 0 – 4000 cm^{-1}) of ninety-three chemicals obtained from NIST (Linstrom and Mallard 2003). Each feature in the absorption spectrum indicates the intensity of light absorbed by a molecule at a particular wavenumber, thus defining a high dimensional odor signal of 4,000 features.

6.4 Modeling Receptor Neuron Convergence

To process high-dimensional experimental data from infrared spectroscopy we adapt the ORN convergence model presented by us earlier (Gutierrez-Osuna 2002; Raman et al., 2006). Briefly, this model is based on three principles: (i) ORNs with similar affinities project onto neighboring GL, (ii) GLs in OB are spatially arranged as a two-dimensional surface, and (iii) neighboring GL tend to respond to similar odors (Meister and Bonhoeffer 2001; Johnson and Leon 2000). Therefore, a natural choice to model the ORN-GL convergence is the self-organizing map (SOM) of Kohonen (1982).

To form a chemotopic mapping, we must first define a selectivity measure upon which IR absorption features can be clustered together. In this work, this is accomplished by treating the IR absorption at a particular wavelength across a set of odorants as an affinity vector:

$$IR_i = \left[IR_i^{O_1}, IR_i^{O_2}, ..., IR_i^{O_C} \right] \tag{6.1}$$

where IR_i^O is the IR absorption at wavenumber i for odor O, and C is the number of odorants (C = 93 in this study).

The convergence model operates as follows. The SOM is presented with a population of IR absorption features (corresponding to each wavenumber), each represented by a vector in C-dimensional affinity space, and trained to model this distribution. Once the SOM is trained, each IR absorption feature is then assigned to the closest SOM node in affinity space, thereby forming a convergence map from which the response of each SOM node is computed as:

$$SOM_j^O = \sum_{i=1}^{N} W_{ij} IR_i^O \tag{6.2}$$

where N is the number of IR features (N = 4,000 in this study), and W_{ij}=1 if IR_i converges to SOM_j and zero otherwise.

To help visualize this model, Fig. 6.2. illustrates a problem with absorption spectra of three odors (labeled as A, B and C). The affinities at different wavenumbers are shown

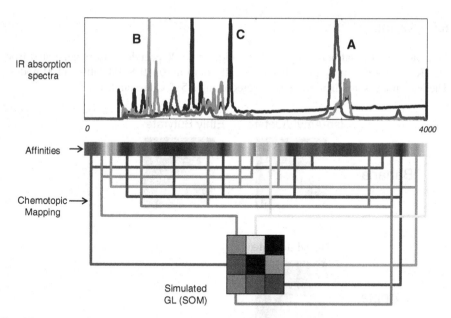

Fig. 6.2. Illustration of chemotopic convergence: the relative response to three analytes (labeled A, B and C) is used to define the wavenumbers' affinities (shown as a colorbar). IR wavenumbers with similar affinities project to the same SOM node as a result of chemotopic convergence. Activity across the SOM lattice can be considered as an artificial odor map.

as a colorbar below the IR spectra[1]. The chemotopic mapping is achieved by assigning features (i.e., IR wavenumbers) with similar affinities to the same SOM node. The activity of the entire SOM lattice is then considered as an artificial odor map.

This convergence model works well when the different sensors are reasonably uncorrelated, since the projection of sensor features across the SOM lattice approximates a uniform distribution, i.e., maximum entropy (Lancet et al. 1993; Laaksonen et al. 2003). Unfortunately, the population of sensors created through IR absorption spectra tends to be over-sampled. As a result, a few SOM nodes tend to receive the majority of input, which capture the "common-mode" response of the sensor, overshadowing the most discriminatory information. To avoid this issue, the activity of each SOM node is normalized by the number of sensor features that converge onto it:

$$SOM_j = \frac{\sum_{i=1}^{N} W_{ij} IR_i}{\sum_{i=1}^{N} W_{ij}} \qquad (6.3)$$

Note that this solution is not driven by biological plausibility but largely by the limitations of the sensors.

---

[1] This is a simplification to illustrate the concept, as the actual affinity space in this case is three dimensional.

6.5 Results

To generate artificial odor maps, a population of 4,000 pseudo-sensors generated from the IR spectrum is projected chemotopically onto a 10x10 SOM lattice (100 nodes). The odor images are then low-pass filtered using a 5x5 Gaussian kernel.

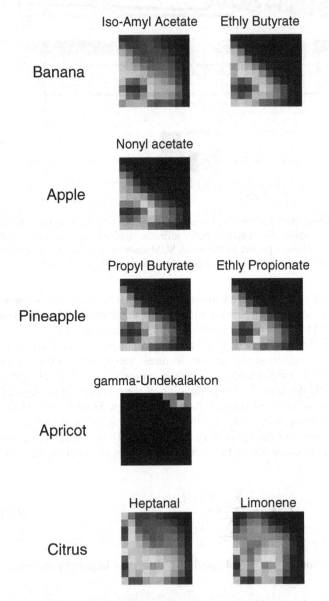

Fig. 6.3. Odor maps generated from the IR spectrum using the chemotopic convergence model for ten different smell percepts: i) banana, ii) pineapple, iii) apple, iv) apricot, v) citrus, vi) nuts, vii) cheese, viii) sweat, ix) minty and x) fat.

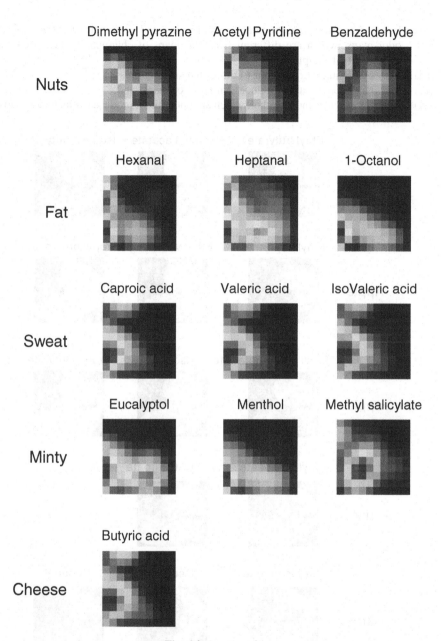

Fig. 6.3. (*continued*)

Fig. 6.3. shows the odor maps for ten different smell percepts[2] from the IR database. The following observations can be made based on the odor images obtained from their IR absorption spectrum:

(i) Esters that smell like tropical fruits (banana and pineapple) produce similar odor maps, which are different from the maps of chemicals with apricots or citrus fruits descriptors,

(ii) *Citrus* odor maps are similar to those that smell *Fatty*,

(iii) *Sweat* and *Cheese* also produce similar odor maps, and,

(iv) Methyl salicylate and Menthol, which are both minty, produce distinct odor maps.

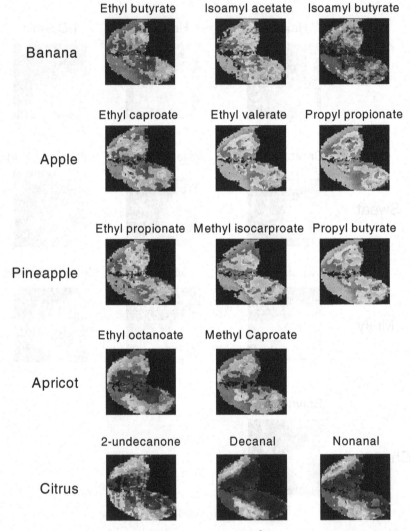

Fig. 6.4. Odor maps obtained in the rat olfactory bulb[3] for the same ten smell percepts

---

[3] A database of odor maps from the rat olfactory bulb is available at http://leonlab.bio.uci.edu/

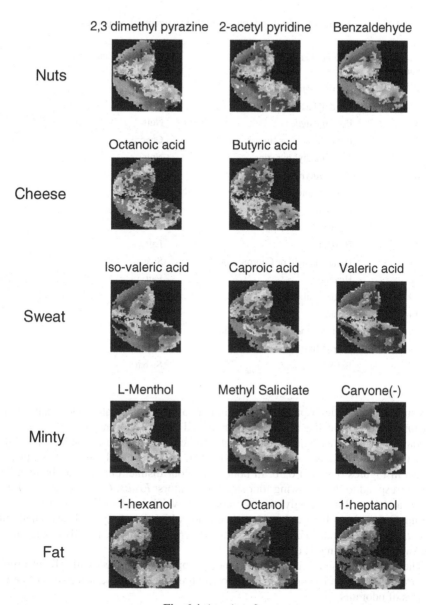

Fig. 6.4. (*continued*)

Spatial odor images for these compounds in the dorsal part of rat OB are shown in Figure 6.4. These odor maps were obtained using optical imaging techniques involving 2-deoxyglucose uptakes in the dorsal part of the rat olfactory bulb (Johnson and Leon 2000). Similar to the images obtained from the IR spectra, esters with tropical fruit smells produce similar activation patterns across the OB, which is different from chemicals with apricot and citrus descriptors. Odor maps for Citrus and Fat descriptors,

Table 6.1. List of odorants and their perceptual properties

Odorant number	Odorant name	Perceptual characteristics
1	Acetyl Pyridine	Nuts
2	Iso-amyl acetate	Fruits
3	Benzaldehyde	Nuts
4	Butanoic acid or Butyric acid	Cheese
5	2,3 Dimethyl pyrazine	Nuts
6	Ethyl Butyrate	Fruits
7	Ethyl Propionate	Fruits
8	Heptanal	Citrus, Fatty
9	Heptanol	Fatty
10	Hexenal	Fatty
11	Hexanoic acid or Caproic acid	Sweat
12	Hexanol	Fatty
13	Methyl Salicylate	Minty
14	Octanol	Fatty
15	Pentanoic acid or Valeric acid	Sweat
16	Propyl Butyrate	Fruits
17	Iso-Valeric acid	Sweat

Sweat and Cheese descriptors overlap similar to the IR-generated odor maps. Minty smelling Methyl salicylate and Menthol produced distinct odor maps.

Hierarchical cluster analysis of the seventeen chemicals, shown in Table 1, present in both the *NIST-IR* dataset and the rat OB image dataset reveal similar groupings, as shown in Figures 6.5a and 6.5b. In both cases, four distinct clusters can be identified that correspond to the following four smell descriptors: *Fruits*, *Cheese* or *Sweat*, *Fat* or *Citrus* and *Nuts*. Interestingly, methyl salicylate, which smells *Minty*, is grouped with the nuts category in both cases. Hexanoic acid, which is a fatty acid that smells like *Sweat*, is grouped under *Fat* or *Citrus* smell descriptor using the rat OB images and in the *Sweat* cluster using IR odor maps.

These results suggest that convergence mapping, combined with IR absorption spectra, may be an appropriate method to capture perceptual characteristics of certain classes of odorants.

6.6 Discussion

What molecular features contribute to the overall percept of smell still remains an open question in olfaction. Three theories have been proposed in an attempt to relate molecular properties of an odorant with its overall quality: vibrational, steric, and odotope theories (Dyson 1938; Moncrieff 1949; Shepherd 1987). The vibrational theory first proposed by Dyson (1938), revisited first by Beck and Miles (1947) and later by Wright (1982) and Turin (1996) (Lefingwell 2002), suggests that vibrations due to stretching and bending of odor molecules are the determinants of odor identity and quality[4]. On the other hand, the steric theory initially put forth by Moncrieff (1949) and later extended by Amoore (1970) (Lefingwell 2002) proposes that odor quality is determined by the shape and size of the odorant molecules. More recently, the odotope or weak shape theory was proposed by Shepherd (1987). According to this theory, odor quality is determined by various molecular features of an odorant (commonly referred to as odotopes), such as carbon chain length or different functional groups.

It is important to note that IR absorption spectroscopy is in fact the basis of the vibrational theory of olfactory reception. This theory has been found to be limited in terms of explaining structure-odor relationships (Rossiter 1996). First, enantiomers, molecules that form non-superimposable mirror images of each other, have identical IR absorption spectrum, yet they can smell differently. e.g., the S- and R- enantiomers

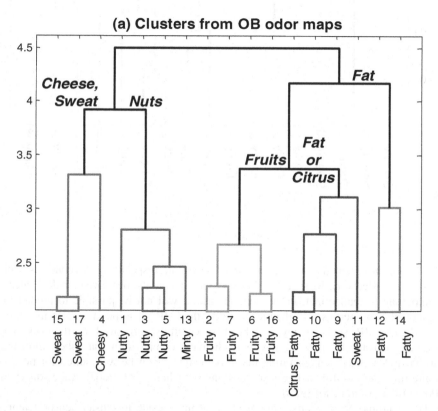

Fig. 6.5. Dendrograms (complete-linkage) revealing similar clusters a) from OB odor maps b) from artificial odor maps formed from their IR absorption spectra. The seventeen common chemicals found in both databases used in this study are listed in Table 6.1.

---

[4] Readers are referred to (Keller and Vosshall 2004) where using psychophysical tests the authors have found that vibrational theory alone cannot explain the overall smell of an odorants.

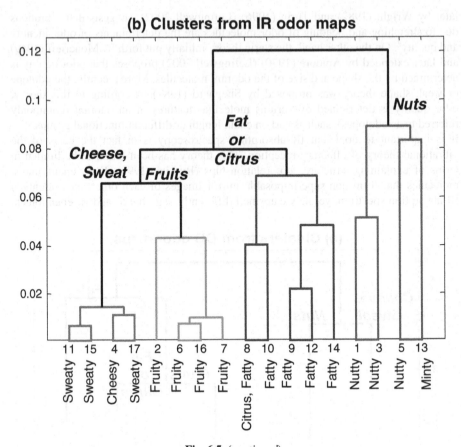

Fig. 6.5. (*continued*)

of carvone have smells of caraway and spearmint, respectively. Second, isotopic substitution affects the IR spectrum but does not change the perceived smell. Hence it is important to realize that, in the general case, it will not be possible to predict the organoleptic properties of chemicals from their IR absorption spectrum alone. Nevertheless, IR spectroscopy has been successfully employed in the food and beverage industry for determining their chemical composition (fat, fiber, moisture, carbohydrates etc.,), demonstrating that this method might be well suited for process monitoring and control in these applications (Li-Chen et al. 2002; Anderson et al. 2002; Osborne and Fearn 1988).

The neuromorphic scheme employed an affinity space to cluster sensor features with similar selectivity. Conventional statistical pattern recognition approaches for clustering operate in the feature space, where each input dimension corresponds to a particular feature (or sensor). Figure 6.6a shows a hypothetical example where multiple samples from two odorants (A, B) have been sampled with a two-sensor array (S1, S2). Samples that belong to the same (odor) class cluster together in feature space, as shown in Fig. 6.6. a. In contrast to feature space, each dimension in affinity space corresponds

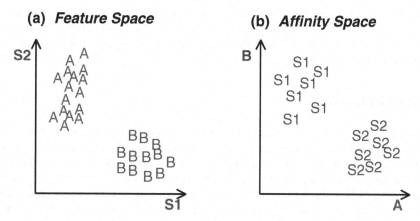

Fig. 6.6. Clustering in feature space and in affinity space. (a) Samples of the same class produce similar response across sensor array, and therefore cluster together in feature space. (b) Features that produce similar response to different odors (classes) cluster together in the affinity (class) space.

to a particular (odor) class. Features that provide similar information regarding the different classes cluster together in the affinity space. As an example, in Fig. 6.6. b all features of type S_1 provide high response to class B and low response to class A; as a result they can be clustered together. In contrast, all features of type S_2 provide high response to class A and low response to class B, therefore form a separate cluster. This basic principle underlies the proposed chemotopic convergence model. A more detailed treatment on the novelty of this approach and its formulation as a dimensionality reduction technique can be obtained from Perera et al. (2006)[5].

6.7 Summary

We have presented a neuromorphic approach for correlating instrumental/sensor data of odorants with their organoleptic properties. This approach comprised of two complementing components: (1) a model of early olfactory processing, which provides odor images that are qualitatively similar to those observed in the OB of animals, and (2) an instrument (IR spectroscopy) that provides high-dimensional data and captures some information consistent with the odotope theory. Our results show that artificial odor maps of chemicals generated from their IR absorption spectra form clusters that match those obtained by clustering the rat OB images of the same set of chemicals. More interestingly, each of these clusters uniquely identified a specific smell descriptor: *Fruits*, *Cheese* or *Sweat*, *Fat* or *Citrus* and *Nuts*. Though encouraging, our results are preliminary at best, as our analysis is limited to those odorants that are common among the NIST and Leon Lab's databases. Further investigations are required to study the relationships among the three representations of an odorant:

---

[5] A related approach to evaluate contribution of a single element in a sensor array has been independently proposed by Niebling and Muller (1995).

stereo-chemical molecular features (Pelosi and Persaud 2000), olfactory bulb images (Johnson and Leon 2000), and organoleptic descriptors (Dravnieks 1985). However, as rightly pointed out by Sell (2006), the complexity of the problem might make such relationships hard to uncover.

Acknowledgments

We are grateful to Alexandre Perera-Lluna for discussions that ultimately led to formulation of chemotopic convergence as a class-space based feature clustering approach. Takao Yamanaka and Agustin Gutierrez-Galvez are also acknowledged for valuable suggestions during this work. We thank Casey Mungle, Josh Hertz, and Jon Evju (PMD, NIST) for their helpful comments on an earlier version of this manuscript. B. R and R. G were supported by the National Science Foundation under CAREER award 9984426/0229598 to R. G during this work.

References

Alkasab, T.K., White, J., Kauer, J.S.: A computational system for simulating and analyzing arrays of biological and artificial chemical sensors. Chemical Senses 27, 261–275 (2002)
Amoore, J.E.: Molecular Basis of Odor. C.C. Thomas, Pub., Springfield, Illinois (1970)
Anderson, S.K., Hansen, P.W., Anderson, H.V.: Vibrational Spectroscopy analysis of dairy products and wine. In: Chalmers, J.M., Griffiths, P.R. (eds.) Handbook of Vibrational Spectroscopy, vol. 5, pp. 3672–3682. Wiley, Chichester (2002)
Axel, R.: The molecular logic of smell. Scientific American 273(4), 154–159 (1995)
Beck, L.H., Miles, W.R.: Some theoretical and experimental relationships between infrared absorption and olfaction. Science 106, 511 (1947)
Buck, L., Axel, R.: A novel multigene family encode odor receptors: a molecular basis for odor recognition. Cell 65(1), 178–187 (1991)
Dyson, G.M.: The scientific basis of odor. Chem. Ind. 57, 647–651 (1938)
Dravnieks, A.: Atlas of Odor Character Profiles, Data Series DS 61, ASTM, Philadelphia, PA (1985)
Firestein, S.: How the olfactory system makes sense of scents. Nature 413, 211–218 (2001)
Friedrich, R.W., Korsching, S.I.: Combinatorial and chemotopic odorant coding in the zebrafish olfactory bulb visualized by optical imaging. Neuron. 18, 737–752 (1997)
Gutierrez-Osuna, R.: Pattern analysis for machine olfaction: A review. IEEE Sensors Journal 2(3), 189–202 (2002)
Gutierrez-Osuna, R.: A self-organizing model of chemotopic convergence for olfactory coding. In: Proc. 2nd Joint EMBS-BMES Conference, Houston, TX, pp. 23–26 (2002)
Hildebrand, J.G., Shepherd, G.M.: Mechanisms of olfactory discrimination: converging principles across phyla. Annual Reviews of Neuroscience 20, 595–631 (1997)
Joerges, J., Küttner, A., Galizia, C.G., Menzel, R.: Representations of odours and odour mixtures visualized in the honeybee brain. Nature 387, 285–288 (1997)
Johnson, B.A., Leon, L.: Modular representation of odorants in the glomerular layer of the rat olfactory bulb and the effects of stimulus concentration. Journal of Comparative Neurology 422, 496–509 (2000)
Keller, A., Vosshall, L.B.: A psychophysical test of the vibration theory of olfaction. Nature Neuroscience 4, 337–338 (2004)

Kohonen, T.: Self-organized formation of topologically correct feature maps. Biological Cybernetics 43, 59–69 (1982)

Laaksonen, J., Koskela, M., Oja, E.: Probability interpretation of distributions on SOM surfaces. In: Proc. Workshop on Self-Organizing Maps (WSOM 2003), Hibikino, Kitakyushu, Japan, pp. 77–82 (2003)

Lancet, D., Sadovsky, E., Seidemann, E.: Probability models for molecular recognitionin biological receptor repertoires: Significance to the olfactory system. PNAS 90, 3715–3719 (1993)

Laurent, G.: A systems perspective on early olfactory coding. Science 286(22), 723–728 (1999)

Leffingwell, J.C.: Olfaction: A review. Leffingwell Reports 2(1), 1–34 (2002) (accessed April 22, 2007), http://www.leffingwell.com

Leon, M., Johnson, B.: Olfactory coding in the mammalian olfactory bulb. Brain Research Review 42, 23–32 (2003)

Li-Chen, E.C.Y., Ismail, A.A., Sedman, J., van de Voort, F.R.: Vibrational spectroscopy of food and food products. In: Chalmers, J.M., Griffiths, P.R. (eds.) Handbook of Vibrational Spectroscopy, vol. 5, pp. 3629–3662. Wiley, Chichester (2002)

Linstrom, P.J., Mallard, W.J.: NIST Chemistry WebBook. NIST Standard Reference Database Number 69, National Institute of Standards and Technology, Gaithersburg (2003) (accessed October 17, 2005), http://webbook.nist.gov

Meister, M., Bonhoeffer, T.: Tuning and topography in an odor map on the rat olfactory bulb. Journal of Neuroscience 21(4), 1351–1360 (2001)

Meurens, M., Yan, S.H.: Applications of infrared spectroscopy in brewing. In: Chalmers, J.M., Griffiths, P.R. (eds.) Handbook of Vibrational Spectroscopy, vol. 5, pp. 3629–3662. Wiley, Chichester (2002)

Moncrieff, R.W.: What is odor. A new theory. Am. Perfumer 54, 453 (1949)

Mori, K., Nagao, H., Yoshihara, Y.: The olfactory bulb: coding and processing of odor molecule information. Science 286, 711–715 (1999)

Niebling, G., Muller, G.: Design of sensor arrays by use of an inverse feature space. Sensors and Actuators B 25(1-3), 781–784 (1995)

Noguiera, F.G., Phelps, D., Gutierrez-Osuna, R.: Development of an infrared absorption spectroscope based on linear variable filters. IEEE Sensors Journal (in press, 2007)

Osborne, B.G., Fearn, T.: Near-Infrared spectroscopy in food analysis. Wiley, New York (1988)

Pearce, T.C.: Computational parallels between the biological olfactory pathway and its analogue The Electronic Nose: Part I. Biological olfaction. BioSystems 41, 43–67 (1997)

Pelosi, P., Persaud, K.C.: Physiological and artificial systems for odour recognition. In: Proc. 2nd Italian Workshop on Chemical Sensors and Biosensors, Rome, Italy, pp. 37–55 (2000)

Perera, A., Yamanaka, T., Gutiérrez-Gálvez, A., Raman, B., Gutiérrez-Osuna, R.: A dimensionality-reduction technique inspired by receptor convergence in the olfactory system. Sensors and Actuators B: Chemical 116(1-2), 17–22 (2006)

Persaud, K.C., Dodd, G.H.: Analysis of discrimination mechanisms of the mammalian olfactory system using a model nose. Nature 299, 352–355 (1982)

Raman, B., Yamanaka, T., Gutierrez-Osuna, R.: Contrast enhancement of gas sensor array patterns with a neurodynamics model of the olfactory bulb. Sensors and Actuators B: Chemical 119(2), 547–555 (2006)

Rossiter, K.J.: Structure-Odor Relationships. Chemical Review 96(8), 3201–3240 (1996)

Sachse, S., Rappert, A., Galizia, G.C.: The spatial representation of chemical structures in the antennal lobe of honeybees: steps towards the olfactory code. European Journal of Neuroscience 11, 3970–3982 (1999)

Schiffman, S.S., Pearce, T.C.: Introduction to olfaction: perception, anatomy, physiology, and molecular biology. In: Pearce, T.C., Schiffman, S.S., Nagle, H.T., Gardner, J.W. (eds.) Handbook of Machine Olfaction: Electronic Nose Technology, pp. 1–32. Wiley-VCH, Weinheim (2003)

Sell, C.S.: On the unpredictability of Odor. In: Angewandte Chemie International Edition, vol. 45(38), pp. 6254–6261. Wiley-VCH, Germany (2006)

Shepherd, G.M.: A molecular vocabulary for olfaction. In: Roper, S.D., Atema, J. (eds.) Olfaction and Taste X, Annals of New York Academy of Sciences, NY, USA, pp. 98–103 (1987)

Turin, L.: A spectroscopic mechanism for primary olfactory reception. Chemical Senses 21(6), 773–791 (1996)

Uchida, N., Takahashi, Y.K., Tanifuji, M., Mori, K.: Odor maps in the mammalian olfactory bulb: domain organization and odorant structural features. Nature Neuroscience 3, 1035–1043 (2000)

Wright, R.H.: The Sense of Smell. CRC Press, Boca Raton (1982)

Zhang, K., Sejnowski, T.J.: Neuronal tuning: to sharpen or broaden. Neural Computation 11, 75–84 (1999)

A Novel Bio-inspired Digital Signal Processing Method for Chemical Sensor Arrays

Eugenio Martinelli[1], Francesca Dini[1], Giorgio Pennazza[1], Maurizio Canosa[1], Arnaldo D'Amico[1,2], and Corrado Di Natale[1]

[1] Department of Electronic Engineering,
  University of Rome "Tor Vergata" via del Politecnico 1; 00133 Roma; Italy
[2] CNR-IA, via del Fosso del Cavaliere; 00133 Roma; Italy

Abstract. In this paper a novel approach to bio-inspired signal processing for artificial olfactory system is proposed. This method is easy to implement in a Field Programmable Gate Array and allows managing a large number of sensor signals using a single chip. It is based on a direct spike conversion of the sensors signals and on the introduction of a digital glomerular signal processing of the spike train. The performance of the method has been compared with standard data analysis also in presence of noisy data.

7.1 Introduction

Chemical sensor arrays were first introduced at mid eighties as a method to counteract the lack of selectivity of gas sensors (Zaromb and Stetter 1984). Persaud argued that besides these practical scopes chemical sensor arrays had a very close resemblance to the mammalian olfactory system opening the way for the advent of artificial olfactory systems (Persaud and Dodd 1984). Moreover with the increase of comprehension of the mechanisms of natural olfaction, it has been possible also to implement in artificial olfactory systems bio-inspired structures not only at high level of abstraction but also as signal processing architectures (Pearce et al. 2001a, Pearce et al. 2001b). Several researchers studied and implemented bio-inspired approached to artificial olfaction. Pearce and co-authors designed and realized a bio-inspired chip attempting to replicate the functions of the natural olfaction from the receptors up to the glomerular level (Pearce et al. 2005). Before these works, Kauer et al. suggested the opportunity to use a bio-inspired neural network for the chemical sensor array data-analysis (White and Kauer 1999). Another interesting approach towards bio-inspired artificial olfaction was presented by Allen and coauthors (Allen et al. 2002, Allen et al. 2004), that studied the possibility to implement a natural olfaction signal processing strategy directly on Field Programmable Gate Arrays (FPGA) using the Address Event Relationship (AER) with different types of digital counters characterized by specific thresholds. In this way, with a single FPGA chip was possible to process more than one thousand sensor signals. Other authors had put their attention to develop biologically derived signal processing and pattern recognition methods for chemical sensor array (Perera et al. 2006, Gutierrez-Galvez et al. 2006, Raman and Gutierrez-Osuna 2006).

A. Gutiérrez and S. Marco (Eds.): Biologically Inspired Signal Processing, SCI 188, pp. 109–120.
springerlink.com   © Springer-Verlag Berlin Heidelberg 2009

Recently we showed that a spike encoding of the sensor signal not only preserves the same information content of the analog sensor signal (Martinelli et al. 2006), but using a software version of the Integrated and Fire circuit and modulating its threshold is possible to improve the recognition obtaining performances that were in some cases better than the standard chemical sensors data treatment.

In this work, starting from these previous results, an approach to chemical sensor signal processing is proposed. This approach is based on an electronic sensor interface that converts analog sensor signal into a train of spikes, with a classic Integrated and Fire circuit, and a novel digital signal processing strategy that allows to process the spikes train incorporating some findings about the glomerular excitation due to simultaneous inputs (Friederich and Stopfer 2001). Results confirm that this method is of certain interest from the pattern recognition point of view.

7.2 Method Description

The proposed approach is a contribution towards the development of artificial systems mimicking as much as possible the natural olfaction. To reach this goal, the artificial olfactory system illustrated in Figure 7.1 is introduced. The system is composed, for each kind of receptor, by three fundamental blocks of the biological olfaction: Olfactive Epithelium, Olfactive Receptor Neuron, Glomerulus. The artificial epithelium is formed by sensor receptors, and in the example here discussed, it is represented by a resistance whose change is produced by the interaction of the sensor with the odorant airborne compounds. The Artificial Olfactive Receptor Neuron (AORN) is defined as a unit transforming the change of resistance into a spikes sequence.

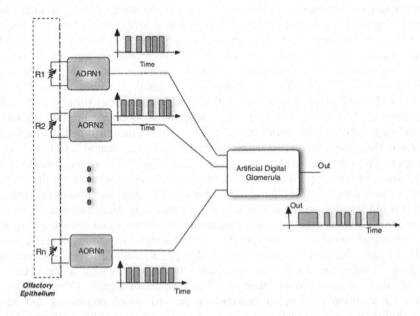

Fig. 7.1. An overview of the proposed bio-inspired architecture of the artificial olfactory

In Figure 7.2, a simple version of an electronic circuit that performs this function is shown. The circuit is based on the well-known Integrated and Fire Circuit where the sensor resistance bridges the op-amp positive input and the voltage supply V_i (Indiveri et al. 2006, Wold et al. 2001).

In this circuit, the relation between resistance change and spikes rate is provided by the following equation:

$$\Delta T = -R_{Sensor} C \cdot \ln\left(1 - \frac{V_{Th}}{V_i}\right)$$ (7.1)

where ΔT is the time interval between two consecutive spikes, C is the capacitance, V_{th} is the voltage threshold and V_i is the voltage applied to the sensor resistance. An inhibitory interval has also been implemented adding a voltage pulse to the threshold voltage immediately after a spike emission. In the following sections, the inhibitory time interval will be maintained constant.

In order to compare the response of the sensor interface with the ORN output spiking rate, they have been considered sensors with a resistance time evolution in presence of a step of chemical stimulus ruled by the equation:

$$R_{sensor}(t) = R_{baseline} + \left[\Delta R(gas, conc) * e^{-\frac{t-t_0}{\tau}}\right] * u(t - t_0)$$ (7.2)

where the ΔR is dependent by the kind and the concentration of the gas. If it has been also considered that all the sensors follow the Langmuir isotherm and the ΔR is correlated with gas concentration through the relation

$$\Delta R = A * \frac{K_1 * conc}{K_2 + K_1 * conc} [\Omega]$$ (7.3)

where K_1 and K_2 are the affinity constants and A is a scale factor.

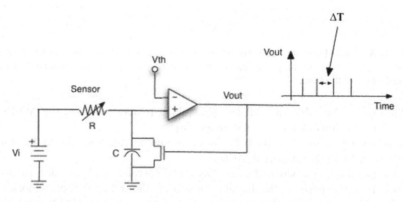

Fig. 7.2. A simple sketch of a Integrated and Fire version where the sensor is represented by the variable resistance R

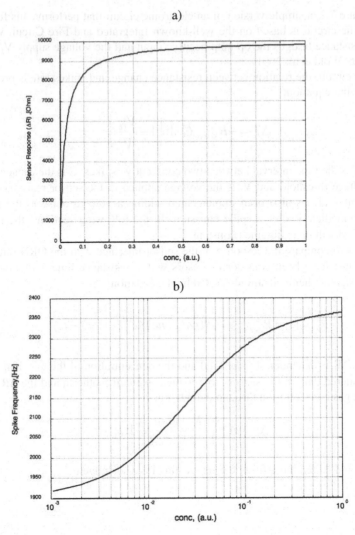

Fig. 7.3. a) A typical Langmuir isotherm behavior of a sensor resistance as a function of gas concentration; b) the spiking frequency rate of the integrated and fire circuit connected with the sensor resistance of the Figure 7.3a

In the Figures 3a and 3b the ΔR and the spiking frequency rate of the AORN of Figure 7.2 as function of the concentration are shown.

It is interesting to remark as the behavior of the spiking rate is similar to the typical behavior of the ORN (Rospars et al. 2000).

The Glomerular block, hereafter called Digital Glomerular block, transfers the spike processing from the analog to the digital context using the spike persistency block (see Figure 7.4). This sub system, that can be also considered as part of AORN, has a similar function of the dendrites-synapses and permits to transform the analog spike into a digital one but also it gives a time length of the spike that can be bigger than the smallest digital event representation (the smallest time interval managed from the clock system). The spike persistency does not produce an overlapping in the spike emission because the time scale of the spiking rate is more than one order of magnitude slower of the clock frequency using in the digital context. Moreover, this block represents an essential characteristic of the proposed architecture because the time length modulation of the spike can simulate the synaptic behavior and it could be used as learning parameter to realize a possible adaptation algorithm.

The digital glomerulus, realized by a combination of logic ports, produces an output spike when a certain number of contemporaneous spikes coming from the ARONs are presented to its input. Figure 7.5 shows the work principle of the digital Glomerulus that gives an output spike with at least three contemporaneous input spikes.

The example illustrates that it is possible to obtain a sequence of spikes that takes into account the number of the input spikes and the time intervals between them. The spike persistency modulation can increase or decrease the contribution of the single AORN in the glomerula output modulating the time length of its pulse. Another

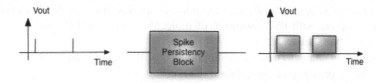

Fig. 7.4. The principle of the Spike Persistency Block. This block provides to enlarge the spike duration.

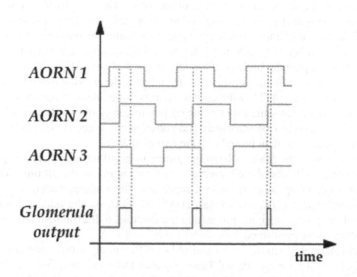

Fig. 7.5. The digital glomerulus functioning. In this case a digital glomerulus realized by an AND logic port with three input. The bigger is the output spike the more synchronized are the input spikes.

feature of the spike persistency is the spike length patterns produced at the glomerula output that express the synchrony of the input AORNs signals.

To test this novel architecture as a tool for classification, a simulated experiment was performed. The case of chemo-resistive sensors was considered because of the simple involved electronics. This class of sensors is rather wide and can include sensors based either on inorganic (e.g. metal-oxide semiconductors) or organic (e.g. conducting polymers) sensitive materials. The concepts here illustrated can be extended, with a proper modification of the AORN architecture, to different kinds of chemical sensors. Actually, the features of the olfactive epithelium define the following structure of the AORN.

7.2.1 Details of Simulated Experiments

A simulated experiment aimed at classifying three odorants was designed to test the consistency of the proposed signal processing method. An array of eight chemically sensitive resistances were considered. The sensor resistance was supposed to change during the exposure to the odorant molecules, according to the equations 2 and 3. At $t=t_0$ the sensor begins to interact with the odorant species. It is worth to mention that Eq. 7.2 is coherent with the adsorption of molecules onto a surface where a limited number of receptors are available.

Each sensor is characterized by different value of $R_{Baseline}$, ΔR and τ, where ΔR and τ change also in presence of the different odorants.

Although in the biological olfaction, the glomerulus has as inputs neurons with the same kind of receptors, in order to test processing capability of the Digital Glomerulus and spike persistency block, it was considered that the neurons connected to the glomerulus could have also different kind of receptors. This simplistic solution allows studying the proposed approach through relatively simple simulations maintaining the validity of the results in more extensive context. Actually the purpose of this work is to increase the number of paradigm of the natural olfaction in the artificial sensorial system trying to implement biological models in a digital context.

To explore the potentialities of this approach, two different kind of digital glomerular blocks were considered. They differ in the number of contemporaneous input spikes necessary to produce the same event at the output, the cases of two (G2) and four (G4) spikes were investigated first separately and subsequently together (G2+G4) to understand the information content brought by the two blocks.

In order to extract the features from the glomerular spiking signal, the output is divided in seven equally distributed intervals over the entire simulation time. The number of spikes occurring in each interval represents the features pattern of the array to the odorant exposure (Rospars et al. 2000). To evaluate the odour identification properties of the proposed signal processing architecture, a discriminant analysis of the feature patterns were performed.

Discriminant analysis indicates a manifold of different algorithms and among them particularly interesting is the Partial Least Squares Discriminant Analysis (PLS-DA) that is a particular way of use of PLS, an algorithm originally developed for quantitative regression (Chicca et al. 2004). PLSDA is here used as simple method to evaluate the proposed signal processing architecture and the Leave One Out Cross Validation (LOOCV) were applied as validation procedure.

To realize the dataset for the data analysis the following characteristics were taken into consideration:

- Three odour classes were defined through the selection of sensors pattern parameters for each class. In the simulation dataset one hundred samples for each class were considered and the same sensor is characterized by the same baseline for all the dataset.
- to study also the noise rejection, a random noise term (**n(t)**) was added to the sensor signal through the following equation

$$R_{Sensor}^{Noisy}(t) = R_{Sensor}(t) \cdot \left(1 + \frac{NoisePercentage}{100} n(t)\right) \tag{7.4}$$

- Olfactory epithelium and AORN were simulated in PSPICE 9.2 environment (Chicca et al. 2004). Spike persistency and digital glomerular block were simulated in Matlab.
- The simulation time was set to 30ms that corresponds to the time necessary to the sensor resistance to steady value. This is to avoid that the output files obtained by PSPICE 9.2 exceeded the file size processable by Matlab (Orcad).
- Each simulation of the entire process has been repeated twenty times and the mean value of the recognition performances was considered.

The performances of the proposed method are compared with the shift between the two steady resistance values (called DR) corresponding to the equilibrium in the cleaning and measuring phase. Each of two equilibrium values has been calculated in presence of noise as the mean value of the ten consecutive samples.

7.3 Results and Discussion

Figure 7.6 shows the results obtained using as inputs of the PLS-DA the time distribution of the output spikes of the G2 glomerulus. The classification rate is represented as function of two parameters: the spike length and the percentage of the added noise. It appears clear as increasing the noise level the performances decrease but it is also interesting to observe how the highest classification rate is always obtained with a spike length interval between 6 and 8. This means that an optimal spike length to maximize the performances exists independently by the noise. The G4 results are shown in the Figure 7.7. The recognition performances behavior is completely different from the G2. In this case also in presence of high level of noise (80% of added noise) the classification rate is always over the 75% but for an high value of the persistency length of the spike.

For the spike persistency value of 4 in presence of 70% of added noise a local minimum in the classification rate is observed. This situation is produced by the data under study and it is not correlated to the proposed approach.

It is interesting to observe that the two kinds of glomerulus have different behaviors in presence of noise and because they represent different projections of information content. In the Figure 7.8 the three dimensional graph of the recognition performances of the model obtained using as inputs both the glomerulus features (G2+G4) is shown.

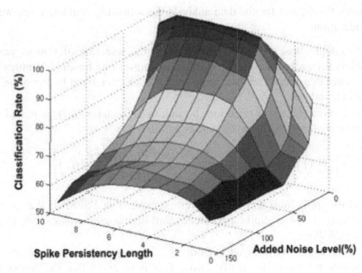

Fig. 7.6. The three dimensional graph of the recognition performances of the PLS-DA model
built using as input the time distribution of the G2 glomerular spikes. The plot is function of the
added noise (%) and the spike persistency length.

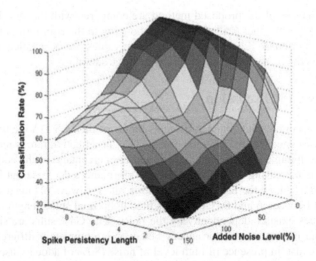

Fig. 7.7. The three dimensional graph of the recognition performances of the PLS-DA model
built using as input the time distribution of G4 (b) glomerular spikes. The plot is function of the
added noise (%) and the spike persistency length.

In this case, the performances remain higher than the two separate glomerula cases. It
is reasonable to suppose that the information contents of the G2 and G4 are not completely overlapped. It is important to remark that also in this case there is an optimal
length of the spike (6-7).

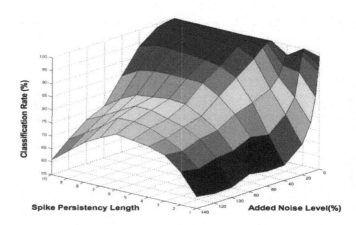

Fig. 7.8. A three dimensional graph of the recognition performances of the PLS-DA model built using as input the time distribution of the G2 and G4 glomerular spikes. The plot is function of the added noise (%) and the spike persistency length.

The Figure 7.9 shows a comparison between the best performances of the Glomerular approaches and those obtained with the correspective DR value. From the graph it is possible to observe an interesting character of the G2 and G4 performances. They have a similar evolution to the DR feature after 20 % of added noise. The model that has as input both G4 and G2 after 70% of added noise results to be the best one. This means that in a very noisy environment, to obtain a reliable value of the resistance shift (DR), it is necessary to calculate the mean value with more than ten samples. On the other hand, the glomerular approach maintains a good discrimination and, as a consequence, a good noise rejection.

After this first phase, it was investigated the recognition power of the different time portion of the glomerular signal. To do this, different PLS-DA models have been built utilizing as input increasing portions of the glomerular signals (the smallest interval considered was 5ms, the bigger one 30ms and the increasing step was 1ms). The feature were calculated counting, as done in the previous case, the number of spikes in each of the seven intervals (equally distributed) that divided the simulation time considered in the analysis.

The results are shown in the Figure 7.10 a-c where the performances of the G2, G4 and G2+G4 are plotted as function of the time interval from the beginning of the measure and spike length. Figure 7.10a shows a independent behavior respect the time intervals. This means that it is necessary only the first 5 ms of simulation in order to extract the total amount of system recognition power. In the case of G4 results (Figure 7.10b), the independent character versus the simulation time showed for G2 is maintained. Anyway, the simulations show a monotonically improvement of the performances increasing of the spike persistency length. Similar graph is obtained using as inputs the features of the two glomeruli (Figure 7.10c). The maximum performances are obtained from the persistency length in the range 4-10. The classification rate is

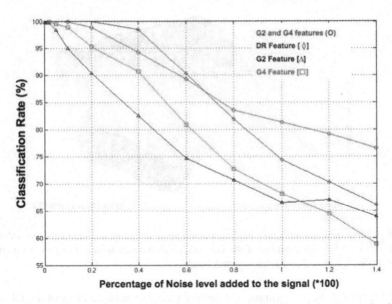

Fig. 7.9. A comparison between the best performances of the DR and the glomerular features has been considered as function of the added noise

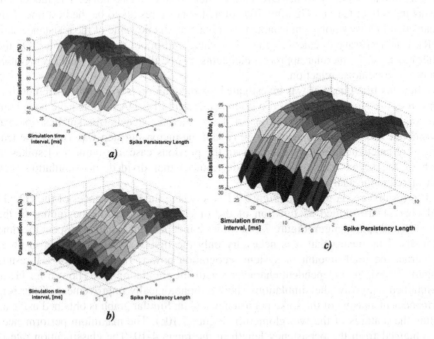

Fig. 7.10. The plot of the PLS_DA model classification rate as function the time portions and spike persistency considered as input the features extracted by G2 (a), G4 (b) and both G2 and G4 (c)

better than the other two cases confirming the assumption related to the different information content brought by the two glomeruli.

It is possible to do some consideration after the simulated experiment results:

- The system architecture is able to maintain the information content of the sensor measurements and the glomerular processing with the spike persistency block gives performances that are comparable to the standard processing algorithm.
- The spike persistency approach allows to find the optimal length of digital spike using a learning rule. In this way it can be possible to implement both supervised and unsupervised algorithm in simple way.
- The proposed structure is modular and then it is possible, easily, to built a system where several ARONs boards are connected to different glomerular processing units.

It is worth to remark that a different number of time intervals in the feature extraction could produce not the same performances. Nevertheless, the study of the optimal interval division of the simulation time from recognition point of view is out of the scope of this work that has as main goal to show the goodness of the proposed architecture.

7.4   Conclusion

An alternative approach to realize a bio-inspired artificial olfactory system is presented. This system is based on the definition of an artificial olfactory Receptor neuron and a Digital Glomerulus block. It is also introduced the spike persistency approach that can allow to transfer in a digital context the glomerular processing. To test the structure a simple simulated experiment is performed and the results showed that this approach it is also interesting not only as a merely bio-inspired method but it can represent an interesting tool to study as a possible useful processing technique. It is also important to remark the flexibility of the spike persistency block that can be also utilized to realize a learning algorithm that uses the persistency lengths as learning parameters. Nevertheless, further studies have to be done in order to investigate other possible digital configurations of the glomerulus block aimed to maximize the information processing and to introduce a modulation of the inhibitory block and to test the structure with real sensors in a experimental application.

References

Allen, J.N., Abdel-Aty-Zohdy, H.S., Ewing, R.L.: Electronic nose inhibition in a spiking neural network for noise cancellation. In: Proceedings of the 2004 IEEE Symposium on Computational Intelligence in Bioinformatics and Computational Biology, CIBCB 2004, pp. 129–133 (2004)

Allen, J.N., Abdel-Aty-Zohdy, H.S., Ewing, R.L., Chang, T.S.: Spiking networks for biochemical detection. In: The 2002 45th Midwest Symposium on Circuits and Systems. MWSCAS 2002, vol. 3, pp. 129–132 (2002)

Chicca, E., Indiveri, G., Douglas, R.J.: An event-based VLSI network of integrate and fire neurons. In: Proceedings of the 2004 International Symposium on Circuits and Systems, 2004. ISCAS 2004, May 23-26, 2004, vol. 5, pp. 357–360 (2004)

Friederich, R.W., Stopfer, M.: Recent dynamics in olfactory population coding. Current Opinion in Neurobiology 11, 468–474 (2001)

Gutierrez-Galvez, A., Gutierrez-Osuna, R.: Increasing the separability of chemosensor array patterns with Hebbian/anti-Hebbian learning. Sensors and Actuators B 116, 29–35 (2006)

Indiveri, G., Chicca, E., Douglas, R.: A VLSI array of low-power spiking neurons and bistable synapses with spike-timing dependent plasticità. IEEE Transactions on Neural Networks 17, 211–221 (2006)

Martinelli, E., D'Amico, A., Di Natale, C.: Spike encoding of artificial olfactory sensor signals. Sensor And Actuators B 119, 234–238 (2006)

Orcad PCB, http://www.orcad.com/pcbdesignerwpspice.aspx

Pearce, T., Mari, C., Covington, J., Tan, F., Gardner, J., Koickal, T., Hamilton, A.: Silicon-based neuromorphic implementation of the olfactory pathway. In: Proceedings of the Second IEEE EMBS Conference on Neural Engineering, Arlington, VA, USA, March 16-19 (2005)

Pearce, T., Verschure, P.F.M.J., White, J., Kauer, J.: Robust stimulus encoding in olfactory processing: Hyperacuity and efficient signal transmission. In: Wermter, S., Austin, J., Willshaw, D. (eds.) Emergent Neural Computational Architectures Based on Neuroscience. LNCS, vol. 2036, pp. 461–479. Springer, Heidelberg (2001)

Pearce, T., Verschure, P.F.M.J., White, J., Kauer, J.: Stimulus encoding during the early stages of olfactory processing: a modeling study using an artificial olfactory system. Neurocomputing 38-40, 299–306 (2001)

Perera, A., Yamanaka, T., Gutiérrez-Gálvez, A., Raman, B., Gutiérrez-Osuna, R.: A dimensionality-reduction technique inspired by receptor convergence in the olfactory system. Sensors and Actuators B 116, 17–22 (2006)

Persaud, K., Dodd, G.: Analysis of discrimination mechanisms in the mammalian olfactory system using a model nose. Nature 299, 352 (1984)

Raman, B., Gutierrez-Osuna, R.: Concentration normalization with a model of gain control in the olfactory bulb. Sensors and Actuators B 116, 36–42 (2006)

Rospars, J.P., Lansky, P., Duchamp-Vriet, P., Duchamp, A.: Spiking versus frequency odorant concentration in olfactory receptor neurons. BioSystems 58, 133–141 (2000)

White, J., Kauer, J.: Odor recognition in an artificial nose by spatiotemporal processing using an olfactory neuronal network. Neurocomputing 26-27, 919–924 (1999)

Wold, S., Sjöström, M., Erikskson, L.: PLS-regression: a basic tool of chemometrics. Chemometrics and Intelligent Laboratory Systems 58, 109–130 (2001)

Zaromb, S., Stetter, J.R.: Theoretical basis for identification and measurement of air contaminants using an array of sensors having partly overlapping selectivities. Sensors and Actuators 6, 225–228 (1984)

8

Monitoring an Odour in the Environment with an Electronic Nose: Requirements for the Signal Processing

A.-C. Romain and J. Nicolas

Department of Environmental Sciences and management, Monitoring unit,
University of Liège, Avenue de Longwy 185, 6700 Arlon, Belgium
acromain@ulg.ac.be

Abstract. Artificial olfaction system (the so-called electronic nose) is a very promising tool to monitor the malodour in the field. Usual odour measurement techniques use human olfaction or analytical techniques. The first category represents the real odour perception but is not applicable to measure in continuous bad odours in the field. The second class of techniques gives the composition of the mixture but not the global information representative of the odour perception. The e-nose has the potentialities to combine "the odour perception" and the "monitoring in the field". However to be able to reach this goal, the signal processing has to be adapted to work in complex environment. We have more than ten years experiments in the measure of environmental malodours in the field and this paper presents the minimal requirements that we consider essential for artificial olfaction system to become successful for this application.

8.1 Introduction

Among all pollution problems, odour annoyance is considered as an important environmental issue since it induces a great number of complaints and conflicts between the residents and the industries. That growing environmental concern has led local authorities to consider odour policies to regulate the odour annoyance. Efforts to manage odour problems and to try to limit the exposure in the neighbourhood are necessary and, of course, the identification and the quantification of odour emission and exposure are very important aspects concerning the compliance with environmental regulation. As the sensitivity to odour involves the highly subjective reaction of individual persons, developing and testing reliable measurement techniques constitute really important challenges when dealing with olfactory pollution.

In this framework, the state-of-the-art generally reports two complementary types of measurement methods: human olfaction methods and analytical techniques (Van Harreveld 2003, Lammers et al. 2004, Stuetz et al. 2001).

Human olfaction measurement considers the odour as a global concept and provides the true dimensions of the human perception. Yet, physiological differences in the smelling of various people often lead to subjective results with large uncertainties. Analytical techniques identify the various volatile compounds involved in the odour and give their chemical concentration. They have better scientific standing than sensory methods. However, the chemical composition of the gas mixture doesn't represent the odour perception.

A. Gutiérrez and S. Marco (Eds.): Biologically Inspired Signal Processing, SCI 188, pp. 121–135.
springerlink.com   © Springer-Verlag Berlin Heidelberg 2009

A third and original concept is promising for measurement of malodour in the environment: the artificial olfaction. The so-called "electronic nose" instrument emerged at the beginning of the nineties thanks to some analogies with the biological sensing system. The instrument, based on non-specific gas chemical sensors array provides a suitable technique for in site monitoring of malodours. Published studies report attractive results (Bourgeois et al. 2003). However a number of limitations are associated with both the properties of chemical sensors and the performances of the signal processing. To meet the requirements of this environmental use of artificial olfaction system, the signal processing method must be simple, but not simplistic, and capable of generalisation. It must be tolerant to hardware weaknesses and adapted for application in the real life.

8.2   Usual Methods to Measure the Odour Pollution

8.2.1   Human Olfaction Measurement

Odour Dimensions

Besides the stimulus, which is a mixture of volatile compounds at given concentrations, the processing of the odour information by the brain is rather complex and leads to a multidimensional sensation. Three main dimensions can be considered: the intensity, or the "strength" of the odour, the quality, or the nature of the odour and the hedonic tone, or the affective reaction to the odour.

Intensity refers to the perceived strength of the odour sensation. To generate such perception, the physical stimulus, i.e. the mixture of odorous molecules, must be detectable. That means that its concentration must exceed a given threshold.

The relationship between the perceived intensity and the concentration of the stimulus, or the odour concentration, is non-linear and depends on the nature of the odorants.

Two famous psychophysical laws express this relationship: the Stevens and the Weber-Fechner laws (Nicolas 2001, Misselbrook et al. 1993, Sperber et al. 2003).

The intensity can be measured by ranking the odour impression on a predetermined scale or by comparison with a series of samples of known concentration of a reference substance (see for example VDI guideline 3882 -1997- Determination of odour intensity).

The notion of odour concentration is based on the works of H. Zwaardenmaker, a Dutch scientist and early investigator of the olfactometry. By definition, the odour concentration, expressed in odorous unit per cubic meter (ou/m3) is the dilution factor of the odour sample in clean air in order to just become odour free, i.e. to reach the perception threshold for an "average" person.

The dynamic olfactometry is the official method by which different dilutions of the gas sample are dynamically presented to trained assessors to determine the odour concentration of the original sample. When the European standard method (EN13725 -European standard "air quality"- Determination of odour concentration by dynamic olfactometry, 2003) is used, the concentration is expressed in $ou_E/m3$ (with the subscript E).

The second dimension of the odour is its quality, expressed in descriptors, or words that describe the smell, such as fruity, woody, sour, pungent (Dravnieks 1992). Contrary to perfumery or oenology, there isn't any standardised list of words to qualify environmental odours and the odour quality is usually associated with its origin, i.e. the emission source.

The third dimension is the hedonic tone, or the category of judgement of the relative like (pleasantness) or unlike (unpleasantness) of the odour. That is an emotional level of reaction which is assessed in accordance with a given category scale, from "extremely unpleasant" to "extremely pleasant" (see for example VDI guideline 3882-2-1994- Determination of hedonic odour tone).

Odour Annoyance Criteria

More specifically for environmental odours, the whole process from the formation of odorants to the complaint action may be complex. It implies many contributing factors and conditions at the different steps.

So, the terminology usually distinguishes the annoyance from the nuisance. Annoyance is simply the negative appraisal of an odour, but nuisance is the cumulative effect on humans, caused by repeated events of annoyance over an extended period of time, that leads to modified or altered behaviour. And, finally, significant odour nuisance may trigger a complaint to a regulatory authority.

Five factors are generally identified in relation to odour impact (Hayes et al. 2006, Schauberger et al. 2001).

Those are the five "FIDOR" (or "FIDOL") factors:

- Frequency, or the number of times an odour is detected over a specific time period,
- Intensity, or the strength of the odour, as defined above,
- Duration, or the length of exposure,
- Offensiveness, or the hedonic tone,
- Receptor, (or "Location") including the physiological, social and economical aspects of the individual perceiving the odour.

Besides the quantitative and qualitative dimensions of the odour, the annoyance is thus chiefly governed by the tolerance and the expectation of the receptor as well as by the time dimension.

The location of the odour event receptor determines how objectionable the odour event is. In some locations certain odours may be more acceptable than in others. For example, breeding smells may be better accepted in rural area or a given industrial smell sometimes does not generate complaints if the company is a job provider for the region.

Concerning the time dimension, frequency and duration may be measured using odour-hours, i.e. the number of hours of odour perception for a given location.

Time variation is actually one of the main specificities of environmental odours, which could invalidate values obtained only by some spot measurements.

For instance, some investigations carried out on 9 landfill sites in Belgium showed that a wide variety of odour emissions are generated by waste. They are conveniently separated into the specific activities, such as active tipping of the waste, its transportation by disposal trucks, its intermediate storage and the handling process after the garbage deposit. On the investigated sites the main odour problem was caused by the handling of the fresh waste, which is an intermittent activity and which makes difficult the sampling of the gas emitted at the landfill working face (Nicolas et al. 2006).

8.2.2 Analytical Techniques

Chemical analysis like gas chromatography coupled to mass spectrometry (GC-MS) carried out in the lab on the basis of samples collected on adsorbent cartridges does not allow monitoring the odour fluctuation in real time.

Alternatively, hand-held specific field detectors allow direct measurement and continuous operation. Particular volatiles, such as hydrogen sulphide, ammonia or total reduced sulphur (TRS), can be continuously recorded and a sudden signal rise can be considered as a sign of the odour emergence for a given source.

However, it is true only if the particular volatiles are correlated to the concentration of the odour of interest. Measuring odours with too specific gas detectors is only suited for emissions with well known gaseous compositions. Moreover, for industrial sites characterised by various gas releases, different types of emission can generate the same signal.

8.3   Artificial Olfaction for Measurement of Odour Pollution

8.3.1   Interest for Environmental Application

Artificial, but biological-inspired odour measurement methods constitute attractive alternatives to the use of human panels to assess the odour pollution in the environment. The evaluation of odour annoyance in the field requires indeed objective approaches and, if possible, means to monitor relevant odorants by a continuous way, even in complex mixtures.

The electronic nose principle exploits that lack of relative specificity to typical gaseous analytes by considering the pattern of sensor signals as global response to the gas mixture. Such bioinspired strategy opened the door in the early eighties to new prospects for those devices that try to mimic the human olfactory system. They include indeed similar corresponding components: the array of chemical sensors, the data processing unit and the pattern recognition engine respectively for the olfactory receptor cells, the olfactory bulb and the brain. Many pattern recognition techniques used to identify the signal patterns apply also bioinspired algorithms, such as artificial neural networks.

One of the main advantages of the electronic nose for malodour measurement is the possibility to use it as a field continuous monitor of odorous emissions (Nicolas and Romain 2000). This technique has probably the best potentialities to answer to the expectations of the various actors of the environmental problems in relation with the odour annoyance.

8.3.2   Stepwise Methodology

To develop an artificial olfaction system and to build the pattern recognition and the odour regression model for this kind of application, we use a stepwise methodology.

The first action to be undertaken is a visit in the field in order to investigate the problem. Then the selection come and the development of the method according to the field criteria and limits. First options are taken and a first prototype is developed and tested in the lab.

A second field campaign is then organised to test the system and to gather some preliminary data to get some proofs before going back a second time in the lab and improving and correcting eventual identified problems.

When the instrument is considered good enough to be carried in the field, then, the learning phase begins. It requires the presence of the operator in the field during several hours and several days to validate each sensor signal variation by sensory observations. Such stage is fundamental to feed the pattern recognition procedures with sufficient data, based on a huge number of various ambiences, to build robust classification and regression models.

A further step in the lab aims at verifying the models in different conditions and with various observation samples. Then a last validation step is carried out in the field.

Such to-and-fros between field and lab result in a final system able to meet field requirements and to bring an optimum solution to the problem.

That quite heavy process must be repeated for each new application, because e-nose is not a universal tool. The only way to guarantee accurate recognition and/or odour concentration prediction is to dedicate the instrument to a specific final utilisation. Dedicated hardware and dedicated signal processing are keys of the e-nose credibility.

8.3.3 Variability of the Operating Conditions

Real life field conditions are never reproducible. For the same site of interest, diverse activities, different process management strategies or various raw material compositions induce highly variable emissions.

As a result, for one odour source, there are many different mixtures of compounds.

So, the various observations belonging to one class of signal patterns are far from being homogeneous.

Figure 8.1 shows two different chromatographic profiles for the same malodorous source.

Other sources of variability are the meteorological conditions. They influence the rate and the composition of the generated odour and affect its dispersion. But they also exert an effect on the gas sensor behaviour (e.g.: signal variation due to humidity change).

Each of those temporal disturbances induces specific signal variation. It contains relevant pieces of information, of different types, and which should not be confused with noise, which is random and erratic. So, to be able to recognise the malodour in the background, among other gaseous emissions, and, for example, to send a warning signal when the odour level rises up above a worrying threshold, or to assess an annoyance zone in the surroundings on the basis of the sensor responses or to use the measured signals to control an odour abatement technique, it is essential to extract just the signal pattern corresponding to the targeted emission and to appraise a significant deviation of the global response from the "normal" margin.

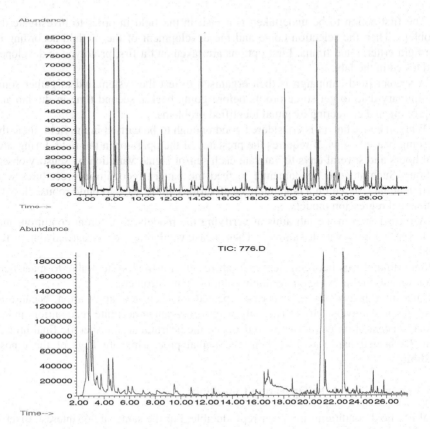

Fig. 8.1. Gas chromatograms obtained by sampling the same odorous source (in a compost shelter) two different days

Figure 8.2 illustrates the time evolution of the conductance of tin oxide gas sensors placed in the ambience of a composting process. Short time signal variations are clearly identified over the noise level and are due to local handlings of compost material. Some other external causes induce diurnal or seasonal variation.

For all those above reasons, trying to classify environmental odour sources in real life with pattern recognition techniques gives rise to a spread of observation points in the different clusters. For the signal processing, that implies to have a great number of samples in order to consider the various conditions and chemical compositions. But that induces a high dispersion of the data of a same class and/or of the same concentration.

The data dispersion is shown in principal component analysis (PCA) scores plot (Figure 8.3). The projection in the two first components plane of the PCA highlights the effect of the variability of the operating conditions on the data.

Such dispersion is the normal image of the large number of samples, collected in many different conditions and with various chemical compositions.

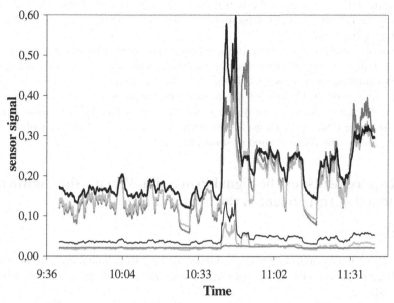

Fig. 8.2. Example of time evolution of 6 tin oxide gas sensors (Figaro™ sensors) in the ambience of a composting plant

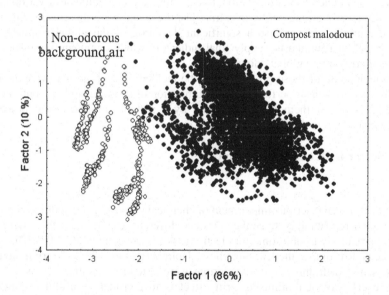

Fig. 8.3. PCA sores plot for measurements with a lab-made e-nose (array of 6 Figaro™ sensors) in a composting plant. The high variance of the data is due to the variability of the operating conditions.

In Figure 8.3, the points in the transitional region between "compost odour" and "odourless background" simply represent less odorous compost ambience or slightly loaded background ambiance.

Moreover, to be able to operate in the environment as an efficient stand-alone instrument, with minimum maintenance, the hardware of the electronic nose has to cope with harsh ambient conditions: dust, humidity, vibration, etc.

And, last but not least, one of the most severe requirements for an instrument aiming at detecting olfactory annoyance is the fact that -contrary to many chemical applications- the final expected useful information is the odour dimension of the gas emission, and not the concentration of some chemicals.

8.4   Requirements for the Signal Processing When Using Artificial Olfaction Instrument in the Environment

8.4.1   Context

As a result of harsh environmental conditions, of hardware limitations and of olfactory pollution specificities, odour real-time monitoring with an electronic nose is thus a real challenge.

The instrument has to cope with several specific drawbacks. It has to automatically compensate the time drift and the influence of ambient parameters, such as temperature or humidity, to filter the unwanted signal variation due to the normal evolution of the background chemical composition, to experiment a huge number of various ambiences during the learning phase, to extract just the signal pattern corresponding to the targeted emission, to appraise a significant deviation of the global response from the "normal" margin and to supply final indicators which aggregate the response of the sensor array into a global odour index.

Although some of those problems may be tackled through an hardware approach (improvement of sensor performances, optimisation of sensor chamber, control of gas line, etc.), most of them must be solved by a suitable on-line signal processing directly embedded in the field instrument.

8.4.2   Sensor Drift and Calibration Gas

Drift

Sensor drift is a first serious impairment of chemical sensors. They alter over time and so have poor repeatability since they produce different responses for the same odour. That is particularly troublesome for electronic noses (Romain et al. 2002). The sensor signals can drift during the learning phase (Holmberg et al. 1997). To try to compensate the sensor drift, three types of solutions were tested for our applications.

The usual way of minimising drift effect is to consider as useful response the difference between the base line, obtained by presenting the sensor array to pure reference air, and the signal obtained after stabilisation in the polluted atmosphere. However, such solution requires operating by cycling between reference air and tainted air, which is not convenient for on-site applications. That requires carrying in the field heavy gas cylinders. Alternatively, generating the reference air by a simple filtering of ambient atmosphere gives rise to only partial drift compensation and to a lack of purity of the reference gas, which increase the data dispersion.

Posterior global drift counteraction algorithms could be applied either for each individual sensor or by correcting the whole pattern, using multivariate methods.

First the main direction of the drift is determined in the first component space of the multivariate method, such as Principal Component Analysis (PCA), or by selecting time as dependant variable of a Partial Least Square regression (PLS). The drift component can then be removed from the sample gas data, correcting thus the final score plot of the multivariate method (Artursson et al. 2000).

Univariate sensor correction gave the best results in our case (table 8.1).

With real-life measurements, it is indeed very difficult to identify a single direction in a multivariate space that is only correlated to sensor drift. So, for each sensor, an individual multiplicative factor was calculated by estimating the drift slope for a standard gas.

Calibration Gas

However, an additional difficulty arises when having to select the adequate calibration gas. There is actually no standard for "compost odour" or "printing house odour".

Criteria to choose a reliable calibration gas are: as simple mixture as possible, with well known compounds, easy to obtain and not too expensive, reproducible, time stable, and, of course, which generates sensor responses similar to the ones obtained with the studied odour.

As it should be difficult to produce stable artificial gas mixture, the best way is to select a single gas.

Figure 8.4 shows drift compensation of a commercial tin oxide sensor (TGS2620, Figaro[TM]) by multiplicative factor estimated from calibration measurements in ethanol.

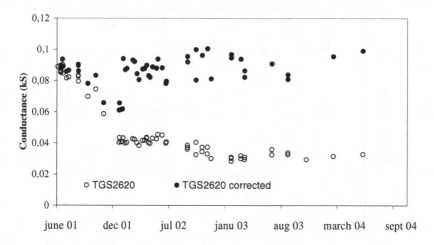

Fig. 8.4. Drift correction of the sensor TGS 2620 by multiplicative factor estimated by ethanol measurements

The method was tested to appreciate the classification performance of an array of tin oxide sensors for two different environmental odours: a compost odour and a printing house odour.

Discriminant Function Analysis was used as classifier, with 5 features (5 sensor conductance values) and 63 observations collected within a 22- month's period. The F-ratio of intergroup/intragroup variances was chosen as classification performance criterion.

Table 1 illustrates the fact that the best classification is achieved with individual sensor correction.

Table 8.1. Evaluation of the DFA classification without correction or with correction models (by the F criterion, F-ratio of intergroup/intragroup variances)

Method	F
No correction	33
Correction by sensor (individual multiplicative factor)	56
Correction of the sensor array "PLS"	26
Correction of the sensor array "PCA"	18

Sensor Replacement

Another frequent problem encountered in the field and particularly in highly polluted atmosphere is a sensor failure or an irreversible sensor poisoning. Clearly, life expectancy of sensors is reduced for real-life operation with respect to clean lab operation. Sensor replacement is generally required to address such issue, but, after replacement, odours should still be recognised without having to recalibrate the whole system.

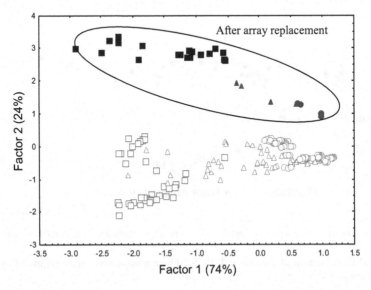

Fig. 8.5. Illustration, by a PCA score plot of the shift of the measurements after the replacement of the TGS sensors

But commercial sensors rarely are reproducible.

Figure 8.5 shows a PCA plot in the plane of the two first components. It concerns 260 observations, 3 classes (ethanol, background air and compost odour), 5 features and a 2-year period. After the replacement of the sensors in the array with the same trade mark references, the previous calibrated model is no longer applicable for the same odorous emissions: all the observation points are shifted to another part of the diagram.

Again, correction routines including algorithms for handling shift related to sensor replacement can be successfully applied. For the above example, illustrated in Figure 8.5, the classification performances were severely reduced after array replacement, the percentages of correct classification were 40%, 100% and 33% respectively for ethanol vapour, background air and compost emission. After individual sensor correction, each classification rate reaches 100 (table 8.2).

Table 8.2. Percentage of correct classification after replacement of sensors with and without correction

Source	$\%^{age}$ correct classification	
	Validation data	
	without correct.	with correct.
ethanol	40	100
air	100	100
compost	33	100

8.4.3 Odour Quantification

One of the final goals of odour monitoring with sensor arrays is the detection of odour emergence in the background. To be able to define some warning threshold or limit values or to assess potential annoyance zones in the surroundings, it is essential to establish a relationship between the e-nose response and the sensory assessment of the odour. But, again, such relationship can not be univocal for a given odour emission, because, as above mentioned, one odorous source corresponds of many different chemical mixtures with variable composition. So, the "x" variable of the relationship, i.e. the e-nose response to volatile compounds, is spread over a large range of possible values. But the "y" branch of the relationship, i.e. the odour sensory assessment, such as the odour concentration measured by dynamic olfactometry, is also affected by large uncertainties. Odour concentration is indeed the result of a rather subjective assessment with a panel of individuals.

Moreover, the gas sensors respond to both odorous and odourless compounds. The condition for the global e-nose signal to be proportional to the "odour" is that the "chemical" concentration and the odour concentration be correlated. That is true for compost emission: there is a quite linear relationship between odour concentration and sensor response (Figure 8.6).

But that is not verified for the odour generated by a printing house (Figure 8.7) because there are a lot of chemical compounds that are not odorous in the ambience.

Odour concentration (uo$_E$/m³)

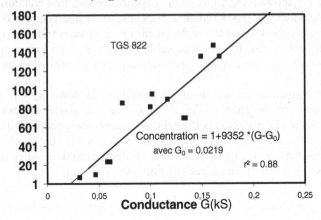

Fig. 8.6. Regression model obtained for compost odour concentration

Odour concentration (uo$_E$/m³)

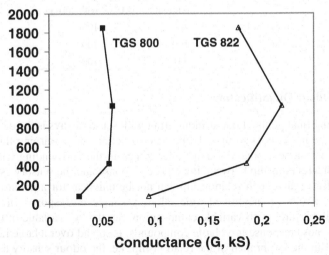

Fig. 8.7. No regression model for a printshop odour

So, in the case of printing house, we cannot use the electronic nose to monitor the odour. But, in the case of compost emission, such monitoring should be possible as long as the relationship between odour and sensor response is kept after eventual sample preparation (eventual pre-concentration, filtering, drying, etc.).

Figure 8.8 shows the evolution of the response of an electronic nose placed in a composting hall. Thanks to a suitable calibration model established with parallel olfactometric measurements, the global response is translated into odour concentration unit (ou$_E$/m³) and could be compared to a warning threshold concentration, e.g. 700 ou$_E$/m³.

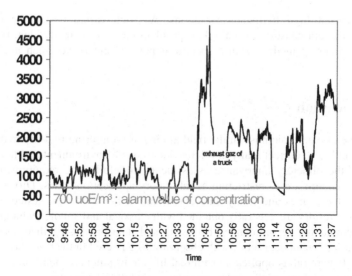

Fig. 8.8. Monitoring of the compost odour concentration

8.4.4 Useful Signal Detection

A last typical example of specific problem when monitoring real-life odours with e-nose is the on-line detection of an abnormal signal emergence.

That issue could be addressed through correct filtering.

Figure 8.9 illustrates such signal processing for an array of sensors placed in the vicinity of settling ponds of a sugar factory. The aim of the filtering process is to keep

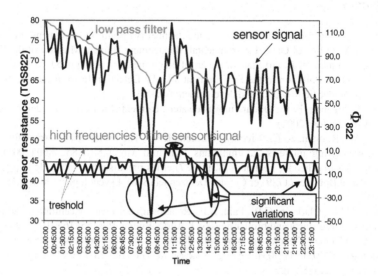

Fig. 8.9. Thank to the filtration of the signal (Φ 822), 4 events are detected (significant variations coupled to an expert system –not shown here-)

only significant high frequency variations. As such variations can be generated by many different causes; the real odour problem must be correctly identified. Combining the filtering method with a specific expert system makes such useful signal detection.

8.5 Conclusion

On-line use of electronic nose in the field aiming at monitoring real-life malodours is feasible, but has to correct many problems due to harsh environment and to the specificities of the olfactory annoyance assessment. Hopefully, the requirements of the final users seldom are as restricting as laboratory applications ones. Different signal processing techniques and many data classification methods were tested, but sufficiently good results are generally obtained with classical and simple techniques. Ideal methods commonly recommended by signal and data processing theories often are well suited for "clean data sets", with observations gathered in the rigorous conditions of the lab, but are rarely applicable with real-life environmental conditions.

The special nature of the environmental malodour monitoring always has to be considered when designing an instrument for such purpose: time variation of the emission, large number of odour categories, sensory dimension of the final variable to assess, lack of standard, etc.

Bio-inspired signal processing development is still necessary, but must be driven by some specific constraints: preliminary field investigation and subsequent on-line validation, even if it is time consuming and not easy, and ability to function in continuous in the field.

References

Van Harreveld, A.P. (Ton): Odor regulation and the history of odor measurement in Europe. In: Ed.: N. a. V. Office of Odor, Environmental Management Bureau, Ministry of the Environment, Government of Japan),State of the art of odour measurement, pp. 54–61 (2003)

Yuwono, L.: Odor Pollution in the Environment and the Detection Instrumentation. the CIGR Journal of Scientific Research and development 6 (2004)

Stuetz, R.M., Frechen, F.-B.: Odours in wastewater treatment: Measuring. Modelling and control, IWA, London (2001)

Bourgeois, W., Romain, A.-C., Nicolas, J., Stuetz, R.M.: The use of sensor arrays for environmental monitoring: interests and limitations. J Environ. Monit. 5(6), 852–860 (2003)

Nicolas, S.: Gustav Theodor Fechner (1801-1887) et les précurseurs français de la psychophysique: Pierre Bouguer (1729) er Charles Delezenne(1828). Psychologie et histoire 2, 86–130 (2001)

Misselbrook, T.H., Clarkson, C.R., Pain, B.F.: Relatioship between concentration and intensity of odours for pig slurry and broiler houses. J. Agric. Engng. Res. 55, 163–169 (1993)

Sperber G, Course: General sensory physiology, Uppsala University (Sweden), Dep. Physiology (2003) (11/03), http://www.neuro.uu.se/fysiologi/gu/nbb/lectures/index.html

Dravnieks, A.: Atlas Of Odor Character Profiles.DS 61, American Society for Testing and Materials (ASTM), Philadelphia (1992)

Hayes, E.T., Curran, T.P., Dodd, V.A.: A dispersion modelling approach to determine the odour impact of intensive poultry production units in Ireland. Bioresource Technology 97(15), 1773–1779 (2006)

Schauberger, G., Piringer, M., Petz, E.: Separation distance to avoid odour nuisance due to livestock calculated by the Austrian odour dispersion model (AODM). Agriculture, Ecosystems and environment 87, 13–28 (2001)

Nicolas, J., Craffe, F., Romain, A.C.: Estimation of odor emission rate from landfill areas using the sniffing team method. Waste Management 26(11), 1259–1269 (2006)

Nicolas, J., Romain, A.C.: Using the classification model of an electronic nose to assign unknown malodours to environmental sources and to monitor them continuously. Sensors and Actuators B: Chemical 69, 366–371 (2000)

Romain, A.-C., André, P., Nicolas, J.: Three years experiment with the same tin oxide sensor arrays for the identification of malodorous sources in the environment. Sensors and Actuators B: Chemical 84, 271–277 (2002)

Holmberg, M., Davide, F.A.M., Di Natale, C., D'amico, A., Winquist, F., Lundstrom, I.: Drift counteraction in odour recognition applications: Lifelong calibration method. Sensors and Actuators B: Chemical 42, 185–194 (1997)

Artursson, T., Eklov, T., Lundström, I., Martersson, P., Sjöström, M., Holmberg, M.: Drift correction for gas sensors using multivariate methods. Journal of chemometrics 14, 711–723 (2000)

9

Multivariate Calibration Model for a Voltammetric Electronic Tongue Based on a Multiple Output Wavelet Neural Network

R. Cartas[1], L. Moreno-Barón[1], A. Merkoçi[1], S. Alegret[1], M. del Valle[1],
J.M. Gutiérrez[2], L. Leija[2], P.R. Hernandez[2], and R. Muñoz[2]

[1] Sensors & Biosensors Group, Department of Chemistry,
Autonomous University of Barcelona, Bellaterra, Catalonia, Spain
Raul.Cartas@campus.uab.es
[2] Bioelectronics Section, Department of Electrical Engineering, CINVESTAV,
Mexico City, Mexico

Abstract. Electronic tongues are bioinspired sensing schemes that employ an array of sensors for analysis, recognition or identification in liquid media. An especially complex case happens when the sensors used are of the voltammetric type, as each sensor in the array yields a 1-dimensional data vector. This work presents the use of a Wavelet Neural Network (WNN) with multiple outputs to model multianalyte quantification from an overlapped voltammetric signal. WNN is implemented with a feedforward multilayer perceptron architecture, whose activation functions in its hidden layer neurons are wavelet functions, in our case, the first derivative of a Gaussian function. The neural network is trained using a backpropagation algorithm, adjusting the connection weights along with the network parameters. The principle is applied to the simultaneous quantification of the oxidizable aminoacids tryptophan, cysteine and tyrosine, from its differential-pulse voltammetric signal. WNN generalization ability was validated with training processes of k-fold cross validation with random selection of the testing set.

9.1 Introduction

An electronic tongue is a chemical analysis system that employs sensors in a novel way, in order to accomplish quantification, classification or identification in liquid media. Conceptually, it relies on the use of a chemical sensor array, with some cross-sensitivity features plus a chemometric processing tool, needed to decode the generated multivariate information. This scheme corresponds to how olfaction and taste senses are organized in animals, allowing for the identification of thousands of different compounds with a reduced number of differentiated receptors, so it is clearly bioinspired. Main types of sensors used in electronic tongues are potentiometric and voltammetric, which yield very different responses. When the nature of the sensors used is voltammetric, a 1-dimensional data vector is generated for each electrode, making extremely complex the chemometric processing of the generated signals. A powerful bioinspired processing tool used with electronic tongues is Artificial Neural Networks (ANNs), although more suited to simpler input information. The use of an ANNs with these signals might then imply some kind of preprocessing stage for data

A. Gutiérrez and S. Marco (Eds.): Biologically Inspired Signal Processing, SCI 188, pp. 137–167.

reduction such as Principal Component Analysis (PCA), Discrete Fourier Transform (DFT) or Discrete Wavelet Transform (DWT) in order to gain advantages in training time, to avoid redundancy in input data and to obtain a model with better generalization ability. Among these techniques, DWT preprocessing has been gaining popularity due to its ability to decompose a signal into two reduced sets of coefficients that separately retain the low and high frequency content of the transformed signal. This variant has been successfully coupled to the work with artificial olfaction systems and also in some chemical analysis situations. In any case, the large effort demanded for optimizing the appropriate set of wavelet coefficients to enter the ANN makes difficult its general use.

Recently, a new class of neural network known as Wavelet Neural Network (WNN) has been proposed. WNN has quickly received favorable opinion due to remarkable results reported for classification, prediction and modeling of different non-linear signals. Very briefly, a WNN is an ANN based on the multilayer perceptron architecture whose sigmoidal activation functions in the hidden neurons are replaced by wavelet functions. Intuitively, it incorporates inside the ANN the task of previously optimizing the wavelet transform for feature extraction of sensor information. In this way, the coupling seems very advantageous for building calibration models, given the setup and fine tuning of the processing conditions are simplified.

Our communication describes the use of a WNN for an electronic tongue employing voltammetric sensors. It starts recalling the parallelisms of electronic tongues with the sense of taste and its physiology. Similarly, the processing of sensorial information through the nervous system is compared to that of ANNs. The backgrounds of wavelet transform are also presented before the conceptual and mathematical principles needed to implement a WNN are described. Finally, we demonstrate an application case in quantitative analysis. The chemical case-study corresponded to the direct multivariate determination of the oxidizable aminoacids tryptophan (Trp), cysteine (Cys) and tyrosine (Tyr), from its differential-pulse voltammetric oxidation signal. We describe the building of the response model from this overlapping signal, its architecture optimization and its verification through k-fold cross validation of input data.

9.2   The Sense of Taste

The human sense of taste occurs as a result of complex chemical analysis starting at a series of chemical active sites called taste buds located within a depression in the tongue. A taste bud is composed of several taste cells (gustatory cells), as shown in Figure 9.1. Each taste bud has a pore that opens out to the surface of the tongue enabling molecules and ions taken into the mouth to reach the taste cells inside.

There are five primary taste sensations including: sweet (carbohydrate based molecules), sour (acidic concentration), salty (sodium chloride), bitter (quinine and other basic functionalities) and umami (salts of glutamic acid). The human tongue does not discriminate every chemical substance composing a flavor; it decomposes the taste of foodstuffs into the five basic taste qualities, instead. A single taste bud contains 50-100 taste cells representing all 5 taste sensation. An adult has about 9000 taste buds.

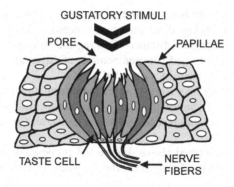

Fig. 9.1. Scheme of a taste bud

Taste cells are stimulated by chemical substances received at its biological membrane. Although the detailed transduction mechanism is not clear, it is known that a nerve fibre connected to several taste cells carries a series of action potentials (electric impulses) in response to chemical stimuli of the taste cells. The connection of each cell is made through a synapse, as usually occurs in transduction of signals in animals; with the difference that a single sensory neuron can be connected to several taste cells of several different taste buds. Each taste bud groups from 50-100 individual sensory cells with a nerve fibre branching out and reaching the nucleus of the solitary fasciculus, where synapses with second order neurons take place. The axons of these neurons extend to the thalamus and form new synapses; finally, neurons from the thalamus send the information to the cerebral cortex to perceive the taste (see Fig. 9.2).

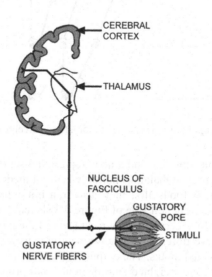

Fig. 9.2. Physiology of the taste sense

The rate of electric impulses transmitted by a nerve fibre increases as the stimuli intensity does. When talking about gustatory receptors, this is accomplished by increasing the concentration of substances forming the flavour. It is worth mentioning that the information transmitted between neurons of different orders (from nucleus to thalamus, for example) is frequency encoded.

A nerve fibre does not necessarily carries the information related to only one taste quality since it does not show nor selective neither specific response (Ogawa et al. 1968). Taste quality is distinguished by using the overall excitation pattern of nerve fibres.

According to the described above, Figure 9.3 proposes a simplified sense-neural scheme showing the sense of taste from the reception of taste substances to the perception in the brain.

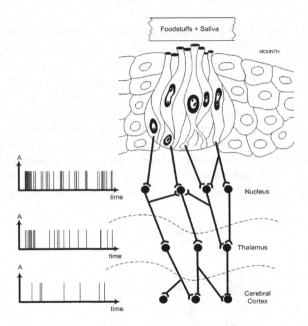

Fig. 9.3. Information path for a taste perception

In Figure 9.3, a connection line and a node represent a set of nerve fibres and neurons, respectively. Neurons in thalamus not only act as a mediator in the transmission of information related to foodstuffs, they also transmit information to neurons in charge of salivary glands. The left side of Figure 9.3 sketches train patterns of electrical impulses travelling by a nerve fibre, at each level of the path. The electrical signal transmitted by each nerve fibre that contacts a neuron is weighted by the synapse and added to the other weighted inputs. The output of the neuron is a train of pulses which frequency is defined by the weighted sum of inputs and probably, by a function applied to this sum, just like in an artificial neuron. This processing of the signal between neurons belonging to different levels is repeated until the information reaches the cerebral cortex, where the taste sensed by the tongue is finally recognized.

One of the goals in gustatory neurobiology is to understand how information about taste stimuli is encoded in neural activity. The taste is defined by two parameters: taste primary and intensity (stimulus concentration). Historically, there have been two major theories of neural coding in the taste system. These are the labeled line (Frank 1973) and across fibre (or neuron) (Erickson et al. 1965) pattern theories. Both of these theories are focused on the spatial representations of neuronal activity.

In the labeled line theory, different taste qualities are encoded by separate groups of cells that respond exclusively, or at least maximally, to a specific quality. In across fibre (or neuron) pattern theory, the relative response magnitude across the population of taste-responsive cells is thought to convey identifying information about taste stimuli. This conceptualization is based on the fact that in multisensitive cells, i.e. cells that respond to more than one taste quality, unambiguous identification of a taste stimulus cannot be gleaned from the simple presence or absence of a response. So, for example, two tastants of different qualities might evoke similar responses rates depending on the particular concentrations at which they are presented. As a result, the relative response magnitude across multiple units may be a better means of stimulus identification.

Both theories depend on a measure of relative response magnitude, most often calculated as the number of spikes (electric impulses) occurring in some arbitrary interval (usually 3-5 s) during which the stimulus is present on the tongue (Di Lorenzo and Lemon 2000). In fact, many of the models presently used to analyze gustatory signals are static in that they use the average neural firing rate as a measure of activity and are unimodal in the sense they are thought to only involve chemosensory information.

More recent investigations of neural coding in the gustatory system have focused on time-dependent patterns of the neuronal response, i.e. temporal coding, as a mechanism of communication in neural circuits (Hallock and Di Lorenzo 2006). In temporal coding, information about taste quality could be conveyed by systematic changes in the firing rate over time within a response, by the timing of spikes during the response, or by the frequency distribution or particular sequence of interspike intervals during the response.

Other researchers (Katz et al. 2002, Jones et al. 2006, Simon et al. 2006) have recently elaborated upon dynamic models of gustatory coding that involve interactions between neurons in single as well as in spatially separate gustatory and somatosensory regions. For dynamic signal processing, these models consider that the information is encoded in time.

9.3 Electronic Tongue

In the field of electrochemical sensors for liquids, there is the recent approach known as electronic tongue, which is inspired on the sense of taste. A widely accepted definition of electronic tongue (Holmberg et al. 2004) entails an analytical instrument comprising an array of non-specific, poorly selective, chemical sensors with cross-sensitivity to different compounds in a solution, and an appropriate chemometric tool for data processing.

The first important part of an electronic tongue is the sensor array; major types of electronic tongues normally use either an array of potentiometric sensors or an array of voltammetric sensors (Gutés et al. 2006), although also other variants have been described, or even their combined use in a hybrid electronic tongue (Winquist et al. 2000). The lack of selectivity in such sensors produces complex cross-response signals that contain information about different compounds plus other features. Due to the lack of selectivity, the second important part in a multisensor approach is the signal processing stage.

The complexity of the signals gathered from the sensor array, perhaps having nonlinear characteristics, and the unknown relationship between the analytes and the sensors' response make the neural networks an ideal candidate for the construction of calibration models.

The reliable performance of electronic tongues in recognition tasks (classification, identification or discrimination) has been demonstrated along the last few years. Potentiometric sensor arrays still are the most widely used type of electronic tongue systems (Ciosek et al. 2004; Legin et al. 1999; Rudnitskaya et al. 2001; Gallardo 2003). Because of their novelty, such systems still need to gain confidence as quality control tools in the food industry, medicine or environmental fields.

An appropriate sensor system will be the first requirement in order to attempt this approach. Additionally, the systems responses must cover the different chemical species and the dynamic range of concentrations expected; these are considered the departure point to build the model needed to create an intelligent system. In this way, the set-up will be able to predict responses of samples not processed initially, as well as to classify them as the human brain does.

The Artificial Neural Network has been the calibration model widely used for electronic tongues. It is possible to stronger relate the electronic tongue with the sense of taste if the theories for encoding and transmitting electric signals through the biological system are considered. As previously mentioned, these theories are labeled line and across fibre (or neuron) pattern theories, and temporal coding.

For electronic tongues using an array of potentiometric sensors, the calibration model based on an ANN corresponds to across neuron pattern theory applied to the biological system.

Finally, when the sensor array is of voltammetric type, signals coming from the sensor array are dynamic (time-varying signals) and commonly non-stationary, which implies that information is time encoded (as stated by recent investigations of neural coding in the gustatory system).

9.4 Artificial Neural Networks

Artificial Neural Networks (ANN) are computational systems that emerged as an attempt to better understand neurobiology and cognitive psychology by means of simplified mathematical models of real neurons (Hassoun 1995; Fine 1999). The initial interest on these systems arose from the hope that they may enable us to increase our knowledge about brain, human cognition, and perception (Garson 2007).

Despite its main objective, biological and knowledge science areas were not the only benefited from neural networks. The success of these systems in tasks such as classification, regression and forecasting attracted the attention of the statistical and engineering communities (Fine 1999; Beale and Jackson 1992), who for the first time have been provided with a tool for building truly non-linear systems with a large number of input variables.

One scientific area in which neural networks have gained popularity is that of the development of systems inspired on olfactory and taste senses. The electronic nose (Gardner and Bartlett 1999), firstly conceived and applied, has gained recognition in fields like food, aroma or medical diagnosis. The electronic tongue (Vlasov and Legin 1998), developed later, has special interest in chemical area for qualitative applications and also for quantifying multiple analytes. Both systems are in continuous development and neural networks have played a key role in its development.

9.4.1 Biological Background

The brain is the organ in charge of constructing our perceptions of the external world, fix our attention and control the machinery of our actions (Kandel et al. 2000). It is composed of about 10^{11} massively connected basic units called neurons of many different types. These units are relatively simple in their morphology, and despite their varieties all of them share the same basic architecture. A neuron (Figure 9.4) has a cell body or soma, where the cell nucleus is located, short nerve fibres branching out of the soma called dendrites that receive most of the incoming signals from many other neurons, and a long tubular fibre called the axon that carries output information away from the soma to its end, where it divides into fine branches to communicate with other neurons at points known as synapses.

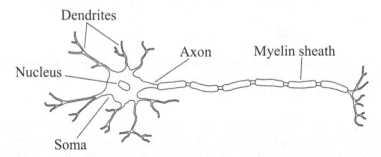

Fig. 9.4. Structure of a multipolar cell neuron, which is the one that predominates in nervous system of vertebrates. It varies in shape, especially in the length of the axon and the number of dendrites.

For a communication between neurons to occur the sender must release substances to excite a receiver and modify its inner potential; if this potential reaches a threshold then an electrical impulse of fixed strength and duration known as action potential is released and travels down the axon to the synapses with other neurons.

Action potentials are the signals by which the brain receives, analyzes and carries information; they are the response of the neurons to many inputs. The information carried by an action potential is determined not by the form of the signal but by the pathway the signal travels in the brain (Kandel et al.2000).

9.4.2 Basic Structure of Feedforward Networks

Artificial neural networks try to mimic the behaviour and structure of the brain. As such, they are constructed by basic processing units with multiple inputs and a single output called artificial neurons.

The earliest neural network is the *perceptron*; it was firstly proposed by McCulloch and Pitts and consists on a single neuron which output is an all-or-nothing value defined by a hard limit function (Hertz et al. 1991). In this structure (sketched on Figure 9.5) the inputs are first weighted by positive or negative values and then fed to the neuron, which sums these products and output a value that depends on if the weighted sum has reached a threshold or not. If the sum of weighted inputs has sur-passed a pre-defined threshold then the neuron fires and takes the activated value (typically 1), otherwise it takes the deactivated value (typically 0). Threshold is com-monly zero but can take any value as long as lies between the minimum and maxi-mum output value. Due to limitations on reflecting the behaviour of a biological neuron the original perceptron is not longer in use today, however, it was the basis for further developments.

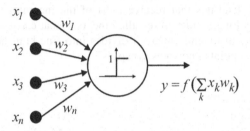

Fig. 9.5. The McCulloch-Pitts' artificial neuron is a bi-state model which output depends on the sum of the weighted inputs. Networks built with this neuron can only solve problems whose solution is defined by a hyperplane.

Larger architectures emerged since then, among them, the *feed-forward multilayer perceptron* (MLP) network has became the most popular network architecture (Hertz et al. 1991). The disposition of neurons in such ANN is quite different from the dispo-sition in the brain; they are disposed in layers with different number of neurons each. Layers are named according to their position in the architecture; an MLP network has an input layer, an output layer and one or more hidden layers between them. Intercon-nection between neurons is accomplished by weighted connections that represent the synaptic efficacy of a biological neuron.

In feed-forward structure (scheme on Figure 9.6), information flows from input to output. Input layer neurons receive the signals coming from outside world, each one compute an output value depending on their activation function and transmit them to all the neurons in the hidden layer they are connected to. These values are weighted by positive or negative values before they reach the inputs of the hidden layer neu-rons. A positive weight means an excitatory contribution and a negative weight means an inhibitory contribution. Hidden layer neurons receive these inputs, add them and

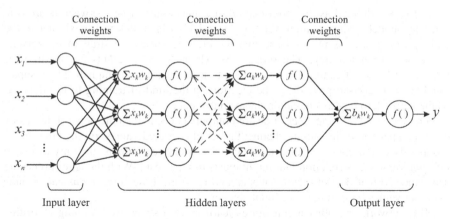

Fig. 9.6. In the MLP network depicted in the figure signal flows from left to right according to the arrows. Each neuron in the hidden or output layer computes a value depending on its activation function and the sum of weighted values coming from a previous layer.

apply a function to this sum to determine the value of their outputs which are passed to output layer neurons or to a second layer of hidden neurons, which in turn do the same actions to determine the value of their outputs.

The way that information flows in a feed-forward network classifies them as hierarchical systems. In such systems, members are categorized by levels, from lowest to highest and they can only communicate from low level to higher but not in the opposite direction. It is worth noting that in a MLP network, input layer neurons do not act as real neurons in the sense that they do not apply an activation function, they act as buffers instead and only distribute the signals coming from outside world to the first hidden layer neurons.

MLP network is not the only kind of network that exists, but it is the one that has shown the best performance for the development of biological inspired systems (Deisingh 2004). Other types of networks are radial basis function (RBF) networks, probabilistic neural network (PNN), generalized regression neural networks (GRNN), linear networks and self organized feature map (SOFM) networks (Freeman and Skapura 1992, Haykin 1999, Hetz et al. 1991, Iyengar et al. 1991); their application depends on the problem we need to solve and their description are out of the scope of this chapter. All these networks can be used to predict values, find patterns in data, filter signals, classify, compress data or model input-output relationships. It is this last application along with its capability to generalize to new situations the reasons for much of the excitement about using neural networks in calibration models of electronic tongues.

9.4.3 Training

Neural networks are processing systems that work by feeding in some variables and get an output as response to these inputs. The accuracy of the desired output depends on how well the network learned the input-output relationship during training.

Learning is defined as the modification of connection weights between neurons to correctly model an input-output relationship (Hebb 1949); this task is similar to the brain learning process in the sense that the brain modifies the strength of synaptic connection when learning. The process by which ANN's weights are modified is known as training. Training can take a long time and can be accomplished by a supervised or unsupervised procedure; the election on the kind of training to use depends on the network to be trained and on the available data.

For supervised training a set of desired outputs is needed, so the network learns by direct comparison of its outputs against the set of expected values; on the other hand, unsupervised training does not count with a set of defined outputs, the only available information is the correlation that might exists in input data. With the latter procedure it is expected that the network creates categories from these correlations and be able to output values according to the input category.

MLP network and other architectures learn using a supervised training algorithm named *error backpropagation*. In this training algorithm input signals are iteratively presented to the network; for each time the input is presented, the network computes its output and compares it against the expected value; next, the difference between them is fed back to the network as an error that is used to adjust the weights of the connections. The aim is to reduce the error and bring the outputs closer to the expected values after each iteration.

Technically speaking, backpropagation is used to calculate the gradient of the network's error with respect to its connection weights. The negative of this gradient and a learning rate parameter are then used in a simple stochastic gradient descent algorithm to adjust the weights and find those values that minimize the error. Large changes in weight values speed up learning and may make it converge quickly to the desired error or overstep the solution; on the other hand, small changes in weight values slower down learning and makes necessary a lot of iterations to converge to the desired error.

9.4.4 Data Selection

Along with the time and effort needed to train a network, a huge input-output data set is required to obtain a reliable generalization model. During training process we must track generalization capabilities of the network, that is, its ability to predict new situations that were not presented during the training process. To accomplish this task the original data set is split into subsets for training, validating and testing the network. Since minimization of error during training does not guarantee that the final network is capable to generalize to new input data we need to check learning progress against an independent data set, the internal validation set.

Each time the weight values are updated during training the performance of the network is evaluated using the validation data set and an error for this set is computed. At the beginning of training process it is common to compute a large output error for both data sets; as training progresses error training decreases and if the network is learning correctly, validation error decreases too. However, if at some point of training validation error stops decreasing or indeed starts to increase then training must be stopped since the network starts to overfit the data. When over-fitting occurs during training it is said that the network is over-learning. On the other hand, if the network can not model the function for which it is being trained then neither training error nor validation error will decrease to the desired value.

Training is a computer controlled process that once started, it might be stopped when either of the next conditions is fulfilled: maximum number of training epochs has elapsed, training and testing errors have reached an acceptable level or there is no more improvement in learning with further iterations.

The third set, the external set is used to check performance of the trained network, and to compare with other configuration or topologies. At the end of training process a well-fit model is the desired result. If the network has properly trained then it will be able to generalize to unknown input data with the same relationship it has learned. This capability is evaluated with the test data set by feeding in these values and computing the output error.

9.5 Wavelet Transform Background

In signal processing, the transform of a signal consists on its manipulation for translating it, from its original domain to another domain, in order to extract its relevant information. The most widespread transform technique is Fourier Transform (FT), which is used for the analysis of periodic signals. FT works by transforming a signal from time domain to frequency domain. It is mathematically defined as

$$F(\omega) = \int_{-\infty}^{\infty} f(t)e^{-j\omega t} dt \qquad (9.1)$$

What FT does is to project the signal $f(t)$ onto the set of sine and cosine basis functions of infinite duration represented by the complex exponential function (Rioul and Vetterli 1991). The transformation (named analysis) is reversible and the recovering of the original function (named synthesis) is done by summing up all the Fourier components multiplied by their corresponding basis function, that is,

$$f(t) = \frac{1}{2\pi} \int_{-\infty}^{\infty} F(\omega)e^{j\omega t} d\omega \qquad (9.2)$$

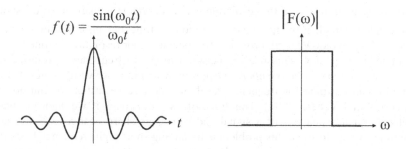

Fig. 9.7. Example of a Fourier transformation. The signal in time domain at left is the sinc function, the magnitude of its Fourier transform is plotted at right. It is noted that F(ω) is band limited.

From the mathematical definition of the Fourier transform, it is advisable that the process determines the frequency content of the signal by obtaining the contribution of each sine and cosine function at each frequency. However, if a transient phenomenon occurred in the signal or it is non-stationary, then the Fourier transform fails either in localizing the anomaly in time domain or representing the signal by the summation of the periodic functions.

A way to solve this bottleneck is by doing the analysis with a piecewise approach known as Short Time Fourier Transform (STFT) or Gabor Transform. The idea behind the process consists on splitting up the signal $f(t)$ into sections and analyse each section separately to find its frequency content (Sarkar and Su 1998). Dividing the signal before its processing helps to localize it in time domain before obtaining the frequency domain information. The division is accomplished by multiplying the original signal with a window function $g(t)$ which most of its energy is centred at time location τ. The goal on using such a weight function for windowing is to avoid spurious frequency components due to abrupt start and end of the window (Alsberg et al. 1997). After multiplication and transformation of a segment of the signal the window function is displaced along time axis to localize a new segment and compute its frequency content again. The process is repeated until the window reaches the end of the signal. The analysis of signals with STFT is done by applying:

$$STFT(\tau, f) = \int_{-\infty}^{\infty} f(t)g(t-\tau)e^{-j\omega t} dt \qquad (9.3)$$

which can be also seen as the convolution of $f(t)e^{-j\omega t}$ with $g(t)$. It is worth to note that if the window function is equal to 1 then we are back to the classic Fourier Transform. Synthesis of signals processed with STFT is done by applying the next formulation

$$f(t) = \frac{1}{2\pi} \int_{-\infty}^{\infty} \int_{-\infty}^{\infty} STFT(\tau, \omega)g(t-\tau)e^{j\omega t} d\tau d\omega \qquad (9.4)$$

STFT maps the original signal into a time-frequency plane (τ, f) which resolution depends critically on the choice of the window function $g(t)$ and its length. Once $g(t)$ has been chosen, then the resolution across the plane is maintained constant. A short window length provides good resolution in time but poor resolution in frequency, a large window length provides the opposite, poor resolution in time but good resolution in frequency. This relation between time and frequency is referred to as *uncertainty principle* or *Heissenberg inequality*, which in other words states that time resolution can be trade for frequency resolution or vice-versa but we can not have both (Rioul and Vetterli 1991). This trade-off is a drawback if we want to obtain a detailed frequency analysis of a signal and isolate its sharp changes (if any) at the same time. A way to solve this problem is by having short high-frequency basis functions and long low-frequency ones. This solution is in fact, achieved by analyzing signals with wavelet transform.

9.5.1 Wavelet Transform

Wavelet transform (WT), a modern tool of applied mathematics, is a signal processing technique that has shown higher performance compared to Fourier transform and
Short Time Fourier Transform in analyzing non-stationary signals. These advantages
are due to its good localization properties in both, the time- and frequency domain.

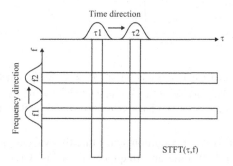

Fig. 9.8. Time-frequency plane described by the STFT. The analysis can be viewed as a series
of FTs defined on windowed segments of the signal (vertical bars) or as a filtering process implemented with a bandpass filter-bank (horizontal bars).

As other processing techniques, WT works by projecting the processed signal
$f(t)$ onto a set of basis functions (this time of finite duration) obtained from a local
wave-like function called mother wavelet $\psi(t)$ by a continuous dilation and translation process (Addison 2002). Dilation, also known as scaling, compresses or stretches
the mother wavelet and limits its bandwidth in the frequency domain, large scaling
values narrow the wavelet and small scaling values broad it. Translation shifts the
wavelet and specifies its position along time axis, positive values shift the wavelet to
the right and negative values shift it to the left. The set of wavelet functions (named
daughter wavelets) and the mother wavelet are related by Eq. 9.5:

$$\psi_{s,\tau}(t)=\frac{1}{\sqrt{s}}\psi\left(\frac{t-\tau}{s}\right), \quad s,\tau \in \Re, \quad s\neq 0 \tag{9.5}$$

For a wave-like function to be considered a wavelet it must fulfil certain mathematical conditions. The most important conditions are *admissibility* and *regularity*
(Mallat 1999). Admissibility condition, expressed as

$$\int_{-\infty}^{\infty}\frac{|\Psi(\omega)|^2}{\omega}d\omega<\infty \quad \Psi(\omega)\text{ is the FT of }\psi(t) \tag{9.6}$$

a) implies that the Fourier transform of the wavelet vanishes at zero frequency, which
is the same as saying that has a zero average in time domain, and b) ensures that the
transformation is invertible, it is said, a signal can be analyzed and synthesized without

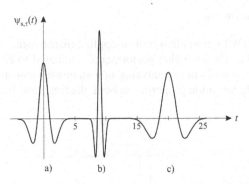

Fig. 9.9. Scaled and translated versions of a wavelet. a) Original wavelet defined by $\psi(t)=(1-t^2)\exp(-t^2/2)$. B) Wavelet scaled by a factor of 0.4 and translated 9 time units to the right. c) Wavelet scaled by a factor of 1.5 and translated 20 time units to the right.

loss of information. Regularity condition states that the wavelet must have smoothness and be concentrated in both time and frequency domains (Graps 1995). For a deeper explanation with rigorous mathematics about wavelets the reader can consult (Goswami and Chan 1999; Bachman et al. 2000; Daubechies 1992) and many others.

The basic idea of WT is to correlate any arbitrary function $f(t)$ with the set of wavelet functions obtained by dilation and translation. A stretched wavelet correlates with low frequency characteristics of the signal, while a compressed wavelet correlates with high frequency characteristics (Blatter 1988). Technically, we can say that scale parameter s relates the spectral content of the function $f(t)$ at a different positions τ (translation parameter, see Figure 9.9). The correlation process described is the Continuous Wavelet Transform (CWT) of a signal, mathematically described as

$$W_f(s,\tau)= \int_{-\infty}^{\infty} f(t)\psi_{s,\tau}^*(t)dt \qquad (9.7)$$

where $\psi_{s,\tau}^*$ is the complex conjugate of $\psi_{s,\tau}$. The wavelet coefficients $W_f(s,\tau)$ indicate how close the signal is to a particular basis function, their values depend on $f(t)$ and in the time region where the energy of $\psi_{s,t}$ is concentrated. Synthesis of wavelet transformation is possible and it consists of summing up all the orthogonal projections of the signal onto the wavelets, that is to say,

$$f(t)=\frac{1}{C_\psi} \int_{-\infty}^{\infty} \int_{-\infty}^{\infty} W_f(s,\tau)\frac{1}{\sqrt{s}}\psi\left(\frac{t-\tau}{s}\right)\frac{dsd\tau}{s^2} \qquad (9.8)$$

Where C_ψ in Eq. 9.8 is a value determined by the admissibility condition.

The CWT, as described in Eq. 9.7, cannot be used in practice because a) the basis functions obtained from the mother wavelet do not form a really orthonormal base, b) translation and scale parameters are continuous variables, which mean that a function $f(t)$ might be decomposed in an infinite number of wavelet functions, and c) there is no analytical solution for most of the wavelet transforms and its numerical calculation is impractical from a computational point of view (Blatter 1988). To overcome these problems, discrete wavelet transform (DWT) has been introduced. For the implementation of DWT translation and dilation parameters must take discrete values instead of continuous. By modifying Eq. 9.5 to:

$$\psi_{j,k}(t) = \frac{1}{\sqrt{s_0^j}} \psi\left(\frac{t - k\tau_0 s_0^j}{s_0^j}\right) \qquad (9.9)$$

we get a new equation for obtaining daughter wavelets at discrete steps. In Eq. 9.9, j and k are integers, s_0^j the dilation factor and τ_0 the translation factor. By doing s_0^j and τ_0 equal to 2 and 1, respectively, we will have dyadic scales and positions that lead to the implementation of efficient algorithms for the processing of discrete signals.

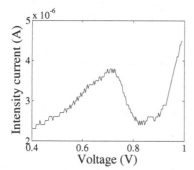 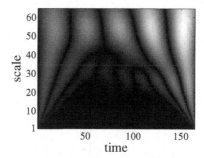

Fig. 9.10. The signal at left was obtained with a voltammetric sensor and was processed with Matlab® to compute the wavelet coefficients plotted at right. Such time-scale plotting of the coefficients is known as scalogram.

$$\psi_{j,k}(t) = 2^{(-j/2)} \psi\left(2^{-j} t - k\right) \qquad (9.10)$$

When discrete wavelets are used to transform a continuous signal, the resulting set of coefficients is called the *wavelet series decomposition*. For this transformation to be useful it must be invertible, and for synthesis to take place, Eq. 9.11 must be satisfied

$$A\|f\|^2 \le \sum_{j,k} \left|\langle f, \psi_{j,k}\rangle\right|^2 \le B\|f\|^2 \qquad A > 0 , \; B < \infty \qquad (9.11)$$

The family of daughter wavelets $\psi_{j,k}$ that satisfy Eq. 9.11 forms a *frame* with frame bounds A and B. When $A = B$ the frame is tight and the analysis and synthesis of the signal can be performed with the same wavelet function, if $A \ne B$ then the wavelet function used for synthesis is different than that used for analysis (Graps

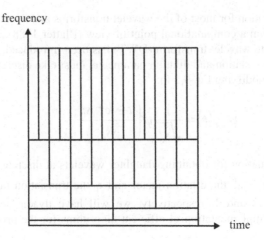

Fig. 9.11. Tiling of the time-frequency plane as defined by the DWT

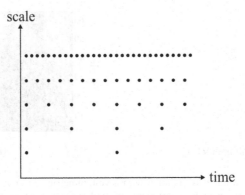

Fig. 9.12. Dyadic sampling of the time-scale plane

1995). For the complete regeneration of the original signal without redundancy it is required that the family of discrete daughter wavelets form an orthonormal basis (Kaiser 1994), that is to say,

$$\int_{-\infty}^{\infty} \psi_{j,k}(t)\psi_{j',k'}(t)dt = \begin{cases} 1 & \text{if } j = j' \text{ and } k = k' \\ 0 & \text{otherwise} \end{cases} \qquad (9.12)$$

This ensures that the information stored in a wavelet coefficient is not repeated elsewhere. In some applications orthogonality is not required since redundancy can help to reduce the sensitivity of noise or improve the shift invariance of the transform. With this supporting theory discrete wavelet transform and inverse discrete wavelet transform can now be defined as

$$X_{j,k} = \int_{-\infty}^{\infty} f(t)\psi_{j,k}(t)dt \qquad (9.13)$$

and

$$f(t)= \sum_{j=-\infty}^{\infty} \sum_{k=-\infty}^{\infty} X_{j,k}\psi_{j,k}(t) \qquad (9.14)$$

The practical implementation of the wavelet transform is not determined yet, since Eqs. 13 and 14 still define a large set of orthonormal wavelet basis. Since admissibility condition states that the behaviour of the Fourier transform of a wavelet is

$$\left| \hat{\Psi}(\omega) \right|^2 \Big|_{\omega=0} = 0 \qquad (9.15)$$

then a wavelet has a band-pass spectrum. This property and the dyadic values of translation and scale parameters benefit the implementation of the discrete wavelet transform as an efficient filtering algorithm. Since compression in time is equivalent to stretching the frequency spectrum and shifting it upwards, then scaling the wavelet by a factor 2^{-1} means to stretch its frequency band-pass characteristic by a factor of 2 and also to shift all frequency components up by the same factor (Aboufadel and Schlicker 1999). By using this approach the set of dyadic scaled wavelets works as a band-pass filters bank that covers the finite spectrum of the processed signal with the spectra of the dilated wavelets. Trying to cover the whole spectrum of the signal down to zero frequency requires an infinite number of filters, since each time a wavelet is stretched in time domain its bandwidth halves. To solve this problem Mallat, one of the mathematician developers of the technique, introduced a so-called scaling function with a low-pass spectrum (Mallat 1989). With this new approach the filter bank used to process any signal is composed by a low-pass filter implemented with the scaling function and a set of band-pass filters implemented with the set of scaled wavelets; as the scaling function cares about the low frequency content of the signal, its width plays an important role because it determines the low-frequency information retained and the number of wavelet coefficients obtained from the decomposition.

The filter bank characteristics of the scaling and wavelet functions are used to implement the DWT in a *subband coding* scheme by using Mallat's pyramidal algorithm (Mallat 1989), as sketched in Figure 9.13. Subband coding consists on analysing a signal by passing it through a filter bank.

This signal processing technique operates over a single discrete signal of length M by decomposing it into orthogonal sub-spaces of length ca. $M/2$. Decomposition is made by applying two digital filters, which involves low-pass (scaling function) and high-pass (wavelet functions) versions along with downsampling. The result of such decomposition is two series of coefficients named approximation coefficients cA_j and detail coefficients cD_j. The cA_j set and cD_j set retain the low-frequency and high-frequency content of the signal, respectively. Moreover, this decomposition can be iteratively applied to approximation coefficients to get components of lower resolution and obtain what is known as multiresolution analysis. Expressions for obtaining cA_j and cD_j sets with the multiresolution analysis are:

$$cA_j(k)= \sum_n h_0(n-2k)cA_{j-1}(n) \qquad (9.16)$$

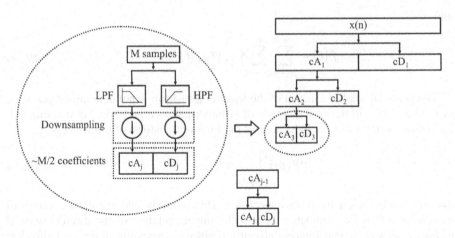

Fig. 9.13. Mallat's pyramidal algorithm used to implement the DWT. Approximation and detail vectors are indicated by cA and cD respectively; subindices denote decomposition level.

$$cD_j(k) = \sum_n h_1(n-2k)cA_{j-1}(n) \qquad (9.17)$$

For each decomposition level '*j*' a faithful reconstruction of the original signal is possible using the inverse discrete wavelet transform (IDWT) and the set of approximation coefficients obtained at level '*j*' altogether with all sets of detail coefficients from level '*j*' until level 1. IDWT is achievable by upsampling the coefficients obtained at level *j* and applying Eq. 9.18:

$$x(n) = cA_0(n) = \sum_k cA_1(k)h_0(n-2k) + \sum_k cD_1(k)h_1(n-2k) \qquad (9.18)$$

In DWT, the digital filters are repeatedly applied to the sets of approximation and detail coefficients until a series of wavelet components obtained at a certain decomposition level is chosen as the result.

9.5.2 Applications Related to the Sensors Field

In the last decade, numerous applications of the WT have been proposed for chemical analysis. One of the main goals in analytical chemistry is to extract useful information from recorded data; however, data gathered from experiments is contaminated with noise. Band-pass filtering behaviour of WT has been successfully applied to the removal of noise or trends, and smoothing (Alsberg et al. 1997).

Data compression is another application of WT that has shown remarkable results (Artursson and Holmberg 2002). The mathematical treatment for data compression by WT is similar to that for denoising and smoothing (Jetter et al. 2000). Chemical data is treated with WT and transformed to the scale-time domain where its spectral content is reduced by eliminating coefficients belonging to high frequency content. Compression with this technique is highly efficient since a one level decomposition and retention of approximation coefficients halves its length. Recovery of the original signal is accomplished by applying inverse WT to the set of coefficients left.

Chromatography, infrared spectroscopy, mass spectrometry, nuclear magnetic resonance spectroscopy, ultraviolet-visible spectroscopy and others have also been benefited from the properties of wavelet processing for data compression, noise removal, base-line correction, zero crossing and regression (Leung et al. 1998).

WT has also been applied to the resolution of overlapping signals in electrochemistry. Frequently, a degree of overlapping among different components is present if different species undergo oxidation or reduction at similar potentials, when using voltammetry. WT has been applied to this situation and has resulted efficient since it offers the advantage of performing data number reduction, feature extraction and noise reduction at the same time (Moreno et al. 2005; Moreno et al. 2006). Recently, Cocchi et al. described a close approach to the work presented here, in which WT is used for feature selection prior to quantitative calibration using ANNs. Compressed voltammetric signals of Pb^{2+}/Tl^+ mixtures were processed with WT and modelled with ANNs (Cocchi et al. 2003; Palacios-Santander et al. 2003).

A field that has clearly made a profit from the advantages of wavelet transform as feature extraction is that of artificial olfaction. There, the pulse response transient corresponding to the injection of a sample can be analyzed dynamically, extracting information with the DWT from the rise and recovery stages, apart from the maximum recorded value, sometimes but not always a steady-state. This strategy was used to identify volatile organic compounds after a PCA analysis using an array of tin-oxide gas microsensors (Distante et al. 2002). The analysis of transient changes has also been extended to the extraction of dynamic response components after a thermal modulation of the working temperature of gas microsensors, aimed to identify between two different gases (CO and NO_2). There, the obtained DWT coefficients were used as input information for a neural network for a classification application (Llobet et al. 2002), or to different chemometric tools for the quantitative determination of both compounds (Ionescu et al. 2002). The same principles were applied shortly after to the correction of humidity interference in a similar system (Ionescu et al. 2003).

9.6 Wavelet Neural Network

The idea of combining wavelets with neural networks resulted in a successful synthesis of theories that generated a new class of networks called Wavelet Neural Network (WNN) (Zhang and Benveniste 1992). This kind of networks use wavelet functions as activation functions in their hidden neurons. Using theoretical features of wavelet transform, methods for building networks can be developed. The first approach to a WNN model makes sense if the inversion formula for the Wavelet Transform (WT) is seen like the sum of products between the wavelet coefficients and the family of daughter wavelets (Akay 1997). This definition established by Strömber replaces the corresponding integrals by a sum, therefore:

$$f(x) = \sum_{s=-\infty}^{\infty} \sum_{t=-\infty}^{\infty} w_{s,t} \psi_{s,t}(x) \qquad (9.19)$$

where $w_{s,t}$ represent the coefficients wavelets of the decomposition of $f(x)$ and $\psi_{s,t}$ the daughter wavelets.

The WNN is based on the similarity found between the inverse WT Strömberg's equation (Eq. 9.19) and a hidden layer in the Multi-Layer Perceptron (MLP) network structure (Meyer 1993). In fact, the wavelet decomposition can be seen like a neuronal network model, where the wavelets are indexed by $i = 1,...,k$ instead of the double index (s,t) that represented the scaled and translated mother wavelet (Figure 9.14).

The compact system of functions located in the hidden layer allows MLPs with only three layers to approximate any arbitrary and continuous function (Hornik 1989, Scarselli 1998). The predetermined precision is defined by the characteristics of the family of functions used as well as by the approach error to reach. In the development of a WNN, a MLP structure with just three layers (input, hidden and output layer) is usually considered, because both the analysis and the implementation are simpler.

Orthogonal wavelets are related with theory of multiresolution analysis and usually cannot be expressed in an informal context; they must fulfill stringent orthogonal conditions, on the other hand, wavelet frames are constructed by simple operations of translation and dilation and are the easiest to use (Akay 1997, Heil 1989, Gutes et al. 2006).

Although many wavelet applications use orthogonal wavelet basis, others work better with redundant wavelet families. The redundant representation offered by wavelet frames has demonstrated to be good both in signal denoising and compaction (Daubechies et al. 1986, 1992).

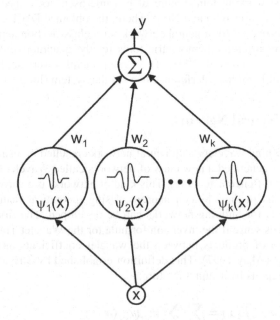

Fig. 9.14. Wavelet expansion observed like a Neural Network

In this way any desired signal $f(x)$ can be approximated by generalizing a linear combination of daughter wavelets $\psi_{s,t}(x)$ derived from its mother wavelet $\psi(x)$. This family of functions is defined by Eq. 9.5 and forms a continuous frame if condition in Eq. 9.11 is fulfilled (Kugarajah and Zhang, 1995).

Nevertheless for multi-variable model's applications it is necessary to use multidimensional wavelets. Families of multidimensional wavelets can be obtained from the product of P monodimensional wavelets, $\psi(a_{ij})$, of the form:

$$\Psi_i(x) = \prod_{j=1}^{P} \psi(a_{ij}) \quad \text{where} \quad a_{ij} = \frac{x - t_{ij}}{s_{ij}} \tag{9.20}$$

where t_i and s_i are the translation and scaling vectors respectively.

9.6.1 WNN Algorithm

The WNN architecture shown in Figure 9.14 corresponds to a feedforward MLP architecture with multiple outputs. The output $y^n(m)$ (where n is an index, not a power) depends on the connection weights $c_i(m)$ between the output of each neuron and the m-th output of the network, the connection weights $w_j(m)$ between the input data and the each output, an offset value $b_0(m)$ useful when adjusting functions that has a mean value other than zero, the n-th input vector x^n and the wavelet function Ψ_i of each neuron. The approximated signal of the model $y^n(m)$ can be represented by Eq. 9.21.

$$y^n(m) = \sum_{i=1}^{K} c_i(m)\Psi_i(x^n) + b_o(m) + \sum_{j=1}^{P} w_j(m)x_j^n \tag{9.21}$$

$$\{i,\ j,\ K,\ P\} \in Z$$

where subindexes i and j stand for the i-th neuron in the hidden layer and the j-th element in the input vector x^n, respectively, K is the number of wavelet neurons and P is the number of elements in input vector x^n.

With this model, a P-dimensional space can be mapped to a m-dimensional space ($R^P \rightarrow R^m$), letting it to predict a value for each output $y^n(m)$ when the n-th vector x^n is input to the trained network (Fig 9.15).

The basic neuron will be a multidimensional wavelet $\Psi_i(x^n)$ which is built using the definition in Eq. 9.20 and in where scaling (s_{ij}) and translation (t_{ij}) coefficients are the adjustable parameters of the i-th wavelet neuron. With this mathematical model for the wavelet neuron the network's outputs becomes a linear combination of several multidimensional wavelets (Zhang and Benveniste 1992, Cannon and Slotine 1995, Mallat 1989, Zhang et al. 1995).

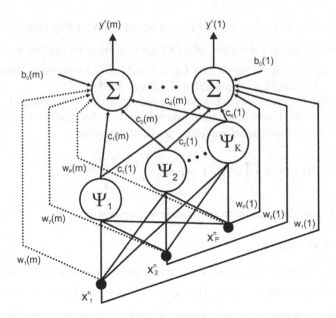

Fig. 9.15. Architecture of the WNN proposed

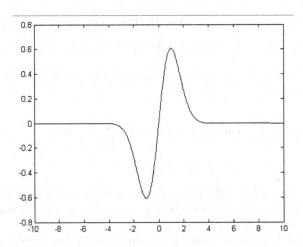

Fig. 9.16. Mother wavelet used as activation function

In the present work, the mother wavelet used as activation function corresponds to the first derivative of a gaussian function (Figure 9.16) defined by $\psi(x) = xe^{-0.5x^2}$, which has demonstrated to be an effective function for the implementation of WNN (Zhang and Benveniste 1992).

9.6.2 Training

Once a network has been structured for a particular application, training of the WNN can proceed. The training process is similar to that of a conventional ANN and in our case consists in the *error backpropagation*. This method, proposed by Rumelhart (Rumelhart et al. 1986), is an iterative algorithm that allows training multilayer networks and helps to determine the neural network parameters. The algorithm looks for the minimum of the error function for the set of training vectors. In our application, weights are updated when all the vectors have been entered to the network. In this form, the training tries to diminish the difference between the outputs of the network and the expected values. The difference is evaluated according to the *Mean Squared Error (MSE)* function defined by Eq. 9.22:

$$J(\Omega) = \frac{1}{2} \sum_{n=1}^{N} \sum_{m=1}^{M} \left(y_{\exp}^{n}(m) - y^{n}(m) \right)^{2} = \frac{1}{2} \sum_{n=1}^{N} \sum_{m=1}^{M} \left(e^{n}(m) \right)^{2} \qquad (9.22)$$

where $y^{n}(m)$ is the m-th output of the network and $y_{\exp}^{n}(m)$ is the m-th real value related to the input vector x^{n}.

Since the proposed model is of multi-variable character, we define:

$$\Omega = \left\{ b_{0}(m), w_{j}(m), c_{i}(m), t_{ij}, s_{ij} \right\} \qquad (9.23)$$

as the set of parameters that will be adjusted during training.

These parameters must change in the direction determined by the negative of the output error function's gradient:

$$-\frac{\partial J}{\partial \Omega} = \frac{1}{N} \sum_{n=1}^{N} \sum_{m=1}^{M} e^{n}(m) \frac{\partial y^{n}(m)}{\partial \Omega} \qquad \text{where} \qquad \frac{\partial y^{n}(m)}{\partial \Omega} = \frac{\partial y}{\partial \Omega} \bigg|_{x=x^{n}} \qquad (9.24)$$

The term $1/N$ averages the output error depending on the number of WNN's outputs.

The changes in network parameters are calculated at each iteration according to $\Delta\Omega = \mu \frac{\partial J}{\partial \Omega}$, where μ is a positive real value known as *learning rate*. With these changes the variables contained in Ω are updated using:

$$\Omega_{new} = \Omega_{old} + \Delta\Omega \qquad (9.25)$$

where Ω_{old} represents the current values , $\Delta\Omega$ represents the changes and Ω_{new} corresponds to the new values after each iteration.

Training is stopped when either the number of epochs has elapsed or the convergence error has been reached.

9.6.3 Initialization of Network Parameters

An important point of the training is the initialization of network parameters, because the convergence of the error depends on them. An initial set of wavelets must be generated at the beginning as well as the weights in the architecture. Considering a range in input vectors defined by the domain $\left[x_{j,\min}, x_{j,\max} \right]$, then the initial values of the i-th neuron for translation and scaling parameters are set to $t_{ij} = 0.5\left(x_{j,\min} + x_{j,\max} \right)$ and $s_{ij} = 0.2\left(x_{j,\max} - x_{j,\min} \right)$, respectively, to guarantee no concentration of wavelets in localities of the input universe

This kind of initialization (Oussar 1998) was shown to be appropriate in a similar model (Gutes et al. 2006). The weights are proposed to have random initial values since its initialization is less critical than translation and scaling variables.

9.7   Case Study in Chemical Sensing

Among the few reports of WNN applied in the chemical area, the modeling and prediction of chemical properties is the main theme, gathering the complexation equilibria of organic compounds with α-cyclodextrins (Guo et al. 1998), the chromatographic retention times of naphtas (Zhang et al. 2001) or the solubility of anthraquinone dyes in supercritical CO_2 (Tabaraki et al. 2006). Much more recent are the applications found in the field of chemical analysis. The first picked reference is the oscillographic chronopotentiometric determination of mixtures of Pb^{2+}, In^{3+} and Zn^{2+} (Zhong et al. 2001), where a discrete WNN was used to build the calibration model. The simultaneous kinetic determination of two species from its spectrophotometric transient record (Ensafi et al. 2007) or the extension of the calibration range for adsorptive stripping voltammetry of Cu^{2+} (Khayamian et al. 2006) are the two most recent applications of WNN.

Authors have applied the principles of the electronic tongue to solve a mixture of three components by direct voltammetric analysis. The approach departs from the overlapped voltammogram, and for the resolution of the three components mixture, a multivariate calibration model is built using WNN. The chemical case in the presented study corresponds to the direct multivariate determination of the oxidizable aminoacids tryptophan (Trp), cysteine (Cys) and tyrosine (Tyr), from the differential-pulse voltammetric signal of a platinum electrode.

For the analysis, three series of synthetic solutions with six concentrations each in the range of [5.0, 35] mM for Cys and Tyr, and [2.0, 21] mM for Trp were prepared. These concentrations were studied at two levels: 10 and 25 mM for Trp and Cys; 5.0 and 34 mM for Tyr. The set of mixture solutions for each analyte was 24, obtained by varying the concentration of the corresponding analyte in its specified range in each of the 4 possible combinations for the level of the two remaining analytes. With this procedure, the total set of mixture solutions and recorded voltammograms for the three analytes is 72.

A commercial potentiostat with a Pt working electrode was used for differential pulse voltammetric measurements. The cell was completed by a second Pt counterelectrode together with an Ag/AgCl reference element. The resulting voltammetric data for each mixture consisted of 164 current values recorded in the range of potentials from 0.4 to 1.0V in steps of 0.00365V. The details of the differential pulse technique were modulation amplitude of 0.025 V, modulation time of 70 ms, and pulse interval of 300 ms. No preconditioning of electrodes was necessary. To generate the series of synthetic samples, microvolumes of each amino acid mixture solution were added to 25 ml of a support electrolyte solution (KCl 0.1M pH 7.5) generating the different sample series.

For the calibration model a three-output WNN with three neurons in its hidden layer was programmed and trained to build the calibration model of our e-tongue. The input layer used 164 neurons, defined as the width of the voltammograms. Error to reach during training was preset to 0.01, the initialization of weights, translation and scaling parameters were according to the descriptions given before and the learning rate was set to 0.001.

The voltammetric matrix constitutes the input data for training and testing the WNN, whereas the concentrations of Trp, Cys and Tyr constitute the targets to be modeled, as sketched on Figure 9.17. For training convenience, the input data and targets were normalized to an interval of [−1, 1] and randomly split into two groups, 54 voltammograms out of the 72 were taken for training and the rest for testing. This process was repeated several times with random subdivision, in what constituted an 18-fold validation scheme.

Training process was successfully accomplished, showing the good characteristics of the WNN for modelling non-linear input-output relationships. Prediction capability was evaluated by constructing comparison graphs of obtained vs. expected concentration values.

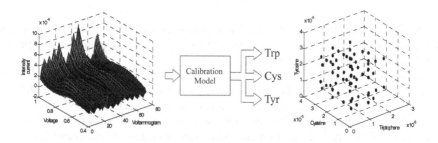

Fig. 9.17. The WNN is in charge of mapping a voltammogram to a point of the three-dimensional space of concentrations

These graphs should be, in the ideal case, the identity line when the scaling is the same for x- and y-axis. This was checked by linear regression analysis, where correlation coefficients should approach 1, and slopes close to 1 along with intercepts close to zero are desirable. Figure 9.18 shows the comparative graphs for the training and testing cases between the expected and predicted concentrations for the three oxidizable aminoacids. For a better reference of the modeling capabilities, each plot has the same scaling for x- and y-axis. As we can see, all points for the training cases lie close

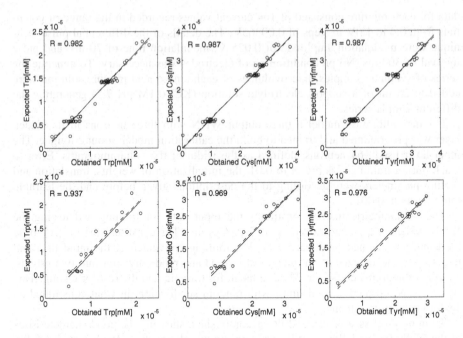

Fig. 9.18. Comparison between the expected results and those obtained with the WNN with 3 hidden neurons and 3 output neurons. The graphs correspond to the three species under study. The dashed line corresponds to ideality (y=x) and the solid line is the regression of the comparison data. Plots at top correspond to training and plots at bottom to testing.

to the identity line, meaning that the value output by the WNN is near the real one, however, for the testing cases the points are more dispersed, although a straight line still can be adjusted to pass through all of them.

Prediction capability of the WNN was evaluated with the testing data set.

The generalization capability of the model was validated with a k-fold validation process. Trainings were run with the initialization parameters described above; training and testing subsets were randomly split and the same size as for the first case. For each model obtained, its prediction capability was evaluated and the correlation factor for training and testing cases were gathered, averaged and summarized in Table 9.1.

Table 9.1. Mean values for correlation coefficients obtained in the k-fold validation process

Aminoacid	Three-output WNN		One-output WNN	
	Training	Testing	Training	Testing
Trp	0.982	0.937	0.984	0.936
Cys	0.987	0.969	0.989	0.980
Tyr	0.987	0.976	0.983	0.966

The performance of the three-output WNN programmed for this case study was compared against a model built with three WNN working in parallel, (one per aminoacid). Good generalization capability of this model has already been reported by (Gutés et al. 2006). In the parallel structure each network had three hidden neurons and one output neuron. Initial conditions of network parameters along with training and testing subsets were the same as for the three-output model. Correlation factors of the k-fold validation process for this model are also summarized in Table 9.1.

Along with the linear regression analysis, the sum of squared errors (SEE) was calculated to give a better understanding of the modeling capabilities of the WNN. These values are contained in Table 9.2.

From results contained in Tables 9.1 and 9.2, we can state that the modeling capabilities of the WNN is nearly the same for both architectures Despite this similarity, the WNN with three-outputs outperforms the approach based on three networks working in parallel since the number of variables needed to train and build the model is reduced by a factor of three. This difference is reflected in the time needed to train the networks and the memory used to store the parameters.

Table 9.2. Mean values for the SEE obtained in the k-fold validation process

Aminoacid	SEE for three-output WNN		SEE for one-output WNN	
	Training	Testing	Training	Testing
Trp	56.17e-12	70.80e-12	55.71e-12	68.79e-12
Cys	131.31e-12	94.43e-12	130.33e-12	91.53e-12
Tyr	9.24e-9	2.46e-9	10.36e-9	2.65e-9

9.8 Conclusions

As it has been shown, the simultaneous quantitative determination of three chemical species has been effectively targeted employing a WNN model. This tool has demonstrated to be a proper multivariate modeling tool for voltammograms, analytical signals with a first degree of complexity that needs their feature extraction prior any modeling or pattern recognition. For its operation, the WNN adjust the parameters for a family of wavelet functions that best fits the shapes and frequencies of sensors' signals. The coupling of wavelet functions with ANNs allows a unique chemometric tool, configurable to many situations and optimizable to the case under study, by the proper selection of the mother wavelet and the topology of the network used.

Acknowledgements

Mobility for researchers between Europe and Latin America was due thanks to the Alfa network project II-0486-FCFA-FCD-FI (European Commission).

References

Aboufadel, E., Schlicker, S.: Discovering wavelets. Wiley, New York (1999)

Addison, P.S.: The illustrated wavelet transform handbook. Institute of Physics Publishing, Bristol (2002)

Akay, M.: Time Frequency and wavelets. In: Akay, M. (ed.) Biomedical Signal Processing: IEEE Press Series in Biomedical Engineering. Wiley—IEEE Press, Piscataway (1997)

Alsberg, B.K., Woodward, A.M., Kell, D.B.: An introduction to wavelet transform for chemometricians: a time-frequency approach. Chemometr. Intell. Lab. Syst. 37, 215–239 (1997)

Artursson, T., Holmberg, M.: Wavelet transform of electronic tongue data. Sens. Actuators B 87, 379–391 (2002)

Bachman, G., Narici, L., Beckenstein, E.: Fourier and wavelet analysis. Springer, New York (2000)

Blatter, C.: Wavelets, a primer. A K Peters Ltd, Natick MA (1988)

Beale, R., Jackson, T.: Neural computing, an introduction. IOP Publishing Ltd., Bristol (1992)

Cannon, M., Slotine, J.E.: Space-frequency localized basis function networks for nonlinear system estimation and control. Neurocomputing 9, 293–342 (1995)

Ciosek, P., Augustyniak, E., Wroblewski, W.: Polymeric membrane ionselective and cross-sensitive electrode-based electronic tongue for qualitative analysis of beverages. Analyst. 129, 639–644 (2004)

Cocchi, M., Hidalgo-Hidalgo-de-Cisneros, J.L., Naranjo-Rodriguez, I., Palacios-Santander, J.M., Seeber, R., Ulrici, A.: Multicomponent analysis of electrochemical signals in the wavelet domain. Talanta 59, 735–749 (2003)

Daubechies, I., Grossmann, A., Meyer, Y.: Painless nonorthogonal expansions. J. Math. Phys. 27, 1271–1283 (1986)

Daubechies, I.: Ten Lectures on wavelets. In: CBMS-NSF Regional Conference Series In Applied Mathematics, Philadelphia, PA. Society for Industrial and Applied Mathematics, vol. 61 (1992)

Deisingh, A.K., Stone, D.C., Thompson, M.: Applications of electronic noses and tongues in food analysis. Int. J. Food. Sci. Technol. 39, 587–604 (2004)

Di Lorenzo, P.M., Lemmon, C.H.: The neural code for taste in the nucleus of the solitary tract of the rat: effects of adaptation. Brain. Res. 852, 383–397 (2000)

Distante, C., Leo, M., Siciliano, P., Persaud, K.C.: On the study of feature extraction methods for an electronic nose. Sens. Actuators B 87, 274–288 (2002)

Ensafi, A.A., Khayamian, T., Tabaraki, R.: Simultaneous kinetic determination of thiocyanate and sulfide using eigenvalue ranking and correlation ranking in principal component-wavelet neural network. Talanta 71, 2021–2028 (2007)

Erickson, R.P., Doetsch, G.S., Marshall, D.A.: The gustatory neural response function. J. Gen. Physiol. 49, 247–263 (1965)

Fine, T.L.: Feedforward neural network methodology. Springer, New York (1999)

Frank, M.: An analysis of hamster afferent taste nerve response functions. J. Gen. Physiol. 61, 588–618 (1973)

Freeman, J.A., Skapura, D.M.: Neural networks: algorithms, applications and programming techniques. Addison-Wesley, Redwood City (1992)

Gallardo, J., Alegret, S., de Roman, M.A., Muñoz, R., Hernandez, P.R., Leija, L., del Valle, M.: Determination of ammonium ion employing an electronic tongue based on potentiometric sensors. Anal. Lett. 36, 2893–2908 (2003)

Gardner, J.W., Bartlett, P.N.: Electronic noses: Principles and Applications. Oxford University Press, Oxford (1999)

Garson, J.: Connectionism. In: Zalta, E.N. (ed.) The Stanford Encyclopaedia of Philosophy (2007), http://plato.stanford.edu/

Goswami, J.C., Chan, A.K.: Fundamentals of wavelets. Wiley, New York (1999)

Graps, A.: An introduction to wavelets. Comput. Sci. Eng. 2, 50–61 (1995)

Guo, Q.X., Liu, L., Cai, W.S., Jiang, Y., Liu, Y.C.: Driving force prediction for inclusion complexation of α-cyclodextrin with benzene derivatives by a wavelet neural network. Chem. Phys. Lett. 290, 514–518 (1998)

Gutés, A., Céspedes, F., Cartas, R., Alegret, S., del Valle, M., Gutierrez, J.M., Muñoz, R.: Multivariate calibration model from overlapping voltammetric signals employing wavelet neural networks. Chemometr. Intell. Lab. Syst. 83, 169–179 (2006)

Hallock, R.M., Di Lorenzo, P.M.: Temporal coding in the gustatory system. Neurosci. Biobehavioral. Rev. 30, 1145–1160 (2006)

Hassoun, M.H.: Fundamentals of artificial neural networks. The MIT Press, Cambridge (1995)

Haykin, S.: Neural networks, a comprehensive foundation. Prentice Hall, Upper Saddle River (1999)

Hebb, D.: The organization of behavior. In: Anderson, A., Rosenfield, E. (eds.) Neurocomputing, foundations of research. The MIT Press, Cambridge (1949)

Heil, C.E., Walnut, D.F.: Continuous and discrete wavelet transforms. SIAM Review 31, 628–666 (1989)

Hertz, J., Krogh, A., Palmer, R.G.: Introduction to the theory of neural computation. Addison-Wesley, Redwood City (1991)

Holmberg, M., Eriksson, M., Krantz-Rülcker, C., Artursson, T., Winquist, F., Lloyd-Spetz, A., Lundström, I.: Second workshop of the second network on artificial olfactory sensing (NOSE II). Sens. Actuators B 101, 213–223 (2004)

Hornik, K.: Multilayer feedforward networks are universal approximators. Neural Networks 2, 359–366 (1989)

Ionescu, R., Llobet, E., Vilanova, X., Brezmes, J., Suegras, J.E., Calderer, J., Correig, X.: Quantitative analysis of nitrogen dioxide in the presence of carbon monoxide using a single tungsten oxide semiconductor sensor and dynamic signal processing. Analyst 127, 1237–1246 (2002)

Ionescu, R., Llobet, E., Brezmes, J., Vilanova, X., Correig, X.: Dealing with humidity n the qualitative analysis of carbon monoxide and nitrogen dioxide using a tungsten trioxide sensor and dynamic signal processing. Sens. Actuators B95, 177–182 (2003)

Iyengar, S.S., Cho, E.C., Phoha, V.V.: Foundations of wavelet neural networks. Chapman & Hall/CRC, Boca Raton (2002)

Jetter, K., Depczynski, U., Molt, K., Niemöller, A.: Principles and applications of wavelet transform to chemometrics. Anal. Chim. Acta. 420, 169–180 (2000)

Jones, L.M., Fontanini, A., Katz, D.B.: Gustatory processing: a dynamic system approach. Current Opinion in Neurobiology 16, 420–428 (2006)

Kaiser, G.: A friendly guide to wavelets. Birkhäuser, Basel (1994)

Kandel, E.R., Schwartz, J.H., Jessell, T.M.: Principles of neural science, 4th edn. McGraw Hill, New York (2000)

Katz, D.B., Nicolelis, M., Simon, S.A.: Gustatory processing is dynamic and distributed. Current Opinion in Neurobiology 12, 448–454 (2002)

Khayamian, T., Ensafi, A.A., Benvidi, A.: Extending the dynamic range of copper determination in differential pulse adsorption cathodic stripping voltammetry using wavelet neural network. Talanta 69, 1176–1181 (2006)

166   R. Cartas et al.

Kugarajah, T., Zhang, Q.: Multidimensional wavelet frames. IEEE Trans. Neural Netw. 6, 1552–1556 (1995)

Legin, A.V., Rudnitskaya, A.M., Vlasov, Y. G., Di Natale, C., D'Amico, A.: The features of the electronic tongue in comparison with the characteristics of the discrete ion-selective sensors. Sens. Actuators B 58, 464–468 (1999)

Leung, A.K., Chau, F., Gao, J.: A review on applications of wavelet techniques in chemical analysis: 1989-1997. Chemometr. Intell. Lab. Syst. 43, 165–184 (1998)

Llobet, E., Brezmes, J., Ionescu, R., Vilanova, X., Al-Khalifa, S., Gardner, J.W., Barsan, N., Correig, X.: Wavelet transform and fuzzy ARTMAP-based pattern recognition for fast gas identification using a micro-hotplate gas sensor. Sens. Actuators B 83, 238–244 (2002)

Mallat, S.: A theory for multiresolution signal decomposition: the wavelet representation. IEEE Trans. Pattern Anal. Mach. Intell. 11, 674–693 (1989)

Mallat, S.: A wavelet tour of signal processing, 2nd edn. Academic Press, San Diego (1999)

Meyer, Y.: Wavelets: Algorithms and Applications. Society for Industrial and Applied Mathematics. SIAM, Philadelphia (1993)

Moreno, L., Cartas, R., Merkoçi, A., Alegret, S., Gutiérrez, J.M., Leija, L., Hernández, P.R., Muñoz, R.: Data Compression for a Voltammetric Electronic Tongue Modelled with Artificial Neural Networks. Anal. Lett. 38, 2189–2206 (2005)

Moreno, L., Cartas, R., Merkoçi, A., Alegret, S., Leija, L., Hernández, P.R., Muñoz, R.: Application of the wavelet transform coupled with artificial neural networks for quantification purposes in a voltammetric electronic tongue. Sens. Actuators B 113, 487–499 (2006)

Ogawa, H., Sato, M., Yamashita, S.: Multiple sensitivity of chorda tympani fibres of the rat and hamster to gustatory and thermal stimuli. J. Physiol 199, 223–240 (1968)

Oussar, Y., Rivals, I., Personnaz, L., Dreyfus, G.: Training wavelet networks for nonlinear dynamic input-output modeling. Neurocomputing 20, 173–188 (1998)

Palacios-Santander, J.M., Jimenez-Jimenez, A., Cubillana-Aguilera, L.M., Naranjo-Rodriguez, I., Hidalgo-Hidalgo-de-Cisneros, J.L.: Use of artificial neural networks, aided by methods to reduce dimensions, to resolve overlapped electrochemical signals. A comparative study including other statistical methods. Microchim. Acta 142, 27–36 (2003)

Rioul, O., Vetterli, M.: Wavelets and signal processing. IEEE SP Magazine 8, 14–38 (1991)

Rudnitskaya, A., Ehlert, A., Legin, A., Vlasov, Y., Büttgenbach, S.: Multisensor system on the basis of an array of non-specific chemical sensors and artificial neural networks for determination of inorganic pollutants in a model groundwater. Talanta 55, 425–431 (2001)

Rumelhart, D.E., Hinton, G.E., Williams, R.J.: Learning internal representations by error propagation. In: Rumelhart, D.E., McClelland, J.L. (eds.) Parallel Distributed Processing: Explorations in the Microstructure of Cognition. Foundations, vol. 1. MIT, Cambridge (1986)

Sarkar, T.K., Su, C.: A tutorial on wavelets from an Electrical Engineering Perspective, Part 2: The Continuous Case. IEEE Antennas and Propagation Magazine 40, 36–49 (1998)

Scarcelli, F., Tsoi, A.C.: Universal Aproximation using Feedforward Neural networks: A survey of some existing methods and some new results. Neural Networks 11, 15–37 (1998)

Simon, S.A., De Araujo, I.E., Gutierrez, R., Nicolelis, M.A.: The neural mechanisms of gustation: a distributed processing code. Nature Rev. Neurosci. 7, 890–901 (2006)

Tabaraki, R., Khayamian, T., Ensafi, A.A.: Wavelet neural network modeling in QSPR for prediction of solubility of 25 anthraquinone dyes at different temperatures and pressure in supercritical carbon dioxide. J. Molec. Graphics Model 25, 46–54 (2006)

Vlasov, Y., Legin, A.A.: Non-selective chemical sensors in analytical chemistry: from electronic nose to electronic tongue. Fresenius J. Anal. Chem. 361, 255–260 (1998)

Winquist, F., Holmin, S., Krants-Rülcker, C., Wide, P., Lundström, I.: A hybrid electronic tongue. Anal. Chim. Acta 406, 147–157 (2000)

Zhang, J., Walter, G.G., Miao, Y., Lee, W.N.W.: Wavelet neural networks for function learning. IEEE Trans. Signal Processing 43, 1485–1497 (1995)

Zhang, Q., Benveniste, A.: Wavelet Networks. IEEE Trans. Neural Netw. 3, 889–898 (1992)

Zhang, X., Oi, J., Zhang, R., Liu, M., Hu, Z., Xue, H., Tao Fan, B.: Prediction of programmed-temperature retention values of naphthas by wavelet neural networks. Comput. Chem. 25, 125–133 (2001)

Zhong, H., Zhang, J., Gao, M., Zheng, J., Li, G., Chen, L.: The discrete wavelet neural network and its application in oscillographic chronopotentiometric determination. Chemometr. Intell. Lab. 59, 67–74 (2001)

Author Index